The Stars Are Not Enough

THE
Stars
ARE NOT
Enough

Scientists—
Their Passions and Professions

Joseph C. Hermanowicz

THE UNIVERSITY OF CHICAGO PRESS ~ CHICAGO AND LONDON

Joseph C. Hermanowicz, a sociologist, lives in Chicago.

The University of Chicago Press, Chicago 60637
The University of Chicago Press, Ltd., London
© 1998 by The University of Chicago
All rights reserved. Published 1998
Printed in the United States of America
07 06 05 04 03 02 01 00 99 98   1 2 3 4 5

ISBN: 0-226-32766-3 (cloth)
ISBN: 0-226-32767-1 (paper)

Library of Congress Cataloging-in-Publication Data

Hermanowicz, Joseph C.
    The stars are not enough : scientists—their passions and
professions / Joseph C. Hermanowicz.
        p.   cm.
    Includes bibliographical references and index.
    ISBN 0-226-32766-3 (cloth : alk. paper). — ISBN 0-226-32767-1
(paperback : alk. paper)
    1. Scientists—Social aspects.   2. Science—Vocational guidance.
I. Title.
Q147.H47   1998
306.4′5—dc21   98-3141
                                                                   CIP

⊗The paper used in this publication meets the minimum requirements of the
American National Standard for Information Sciences—Permanence of Paper
for Printed Library Materials, ANSI z39.48-1992.

for my father
Henry J. Hermanowicz
in continual and loving celebration

CONTENTS

FIGURES AND TABLES

Figures

Tables

PREFACE

The scientists described in these pages, like the people one would encounter in nearly every other profession, are of many stripes. Some are old, some young, others middle-aged. They have careers that have remained steady, careers that have waned, and careers that have "taken off." Looking at these variations, this book examines the perspectives that scientists have come to use to view their careers: where they are, where they have been, where they see themselves headed. These perspectives provide a multicolored lens through which we can see not only what drives people but also what sometimes fails to drive them. More simply, this book asks and explores *what makes people tick.*

This work stands apart from much of what one finds in the sociology of science. On the one hand, it has an unquestionable home in the sociology of science because it is about scientists and the institutional worlds in which they work. On the other hand, it also finds a legitimate home among studies of the human life course, a field that has, until now, been wholly divorced from studies of science. Yet even though previously far removed from one another, these specialties work well together in productive ways and hold true potential for opening interesting new territory—the study of careers in context—as I hope this book will show. The result of this bifocal perspective is an account of how real-life scientific careers, with their real-life drama, take shape and unfold.

Those familiar with the current (and even the old) landscape of the sociology of science will quickly recognize that this work points in a different direction. Currently, many sociologists of science are preoccupied with the "social construction of scientific knowledge." Encompassing diverse approaches, "constructionism" consists generally of a line of inquiry that is concerned with the social conditions and processes that selectively shape what comes to count as "scientific" knowledge. Yet the construction of scientific knowledge need not be the only concern for the sociology of science. To date, the social construction tradition has focused on *knowledge,* whereas I will focus on *scientists*—the people who produce the knowledge, regardless of how selective that knowledge may be. This is a study not about the construction of *knowledge* in science but about the construction of *careers* in science. In this book, I propose that we venture in a new direction by seeing how careers unfold within the social environments in which scientists work. What we gain, in addition to a new perspective, is an in-depth look at scientists as people.

Although my undertaking differs in many respects from the studies produced by those who work in the "constructionist" vein, I hasten to add that we undoubtedly have points in common. Though one may look at knowledge and another at careers, both views inquire about the socially situated circumstances that shape what we call "reality." Those readers interested in how people create meaning about the natural world will likely find interest in a study of how people create meaning about themselves and their place in that world.

In making clear that this study is about scientists and not scientific knowledge, I realize that there are those who might wonder whether this work is simply a return to an older line of inquiry—the one spawned largely by Robert K. Merton and his students, a field that has come to be called the Columbia tradition in the sociology of science. Although the sociology of scien*tists* was a central interest of the Merton school, this tradition concentrated on matters such as performance patterns, reward systems, and the nature of stratification in science; it said little about subjects that I will say much about here: identity and the symbolic meanings of work. Moreover, the Columbia tradition was heavily quantitative and rather distant from the interpretative style of narrative analysis that I engage in. More important, by bringing a life course perspective to careers, I embark on a study of scientists in ways that the "Mertonians" never attempted.

This book speaks to several audiences. I have not tried to be all things to all specialties; accordingly, the work speaks to these audiences at admittedly different decibels. Because this book deals explicitly with how people experience and publicly present the unfolding of their lives, it is intended to speak loudly to those interested in life course studies. Because it looks in particular at careers, this work speaks to those interested in occupations. And because it is centrally concerned with how people create self-identities as their careers unfold, it speaks to those interested in social psychology.

In addition, this book speaks to those interested in education. In building on studies of the life course, I have situated my subjects in—rather than detached them from—the sociocultural contexts in which they work: three distinct types of universities in the system of U.S. higher education. The ways in which these different settings bear on the career, giving it shape and helping build narratives that account for the courses that individuals travel, form a major part of my project. I deal at length with the structure and culture of schools. This work thus addresses the sociology of education, particularly that segment concerned with the study of universities.

Because scientists are my subjects, and because much of the book is about career imagery, aspirations, fears, and anxieties, this book also has an audience in working scientists (and those like them) who are interested in their careers and in the careers of others.

Finally, this book has a more general target, which gives me the occasion to turn to what makes *this* book tick, from start to finish. In doing so, I am called to answer the question of why *anyone,* in the academy or outside it, might want to journey alongside careers in science, this esoteric set of fields that often seems so remote to so many. The answer is that careers in science tell a general social story about all people whose lives are lived with the thought of, if not the active desire for, great accomplishment, the achievement of a kind of immortality.

The scientist, like the artist or the athlete, follows a pantheon of immortals—figures who have achieved a place in history through exemplary performance. The achievements of these figures set a competitive standard for all those who embark on a given professional path. Who aspires to be a mediocre Johnny Unitas, a second-rate Beethoven, or a watered-down Newton? A desire for greatness—an imagined possibility of what one can become—ignites a career and often sustains it. This is as true for scientists as it is for artists, architects, astronauts, physicians, lawyers, composers, poets, politicians, or athletes—all occupations whose members are judged ultimately by lofty standards of originality, invention, and heroic accomplishment.

But so few achieve greatness; not many more achieve anything that comes close to it. Only after having lived a real-life career do people become resigned to a status far outside the secular pantheon of their field. *The Stars Are Not Enough* tells us not only about the lives of scientists but about the lives of all people whose calling sets them on heroic quests and also about the wider culture that highly values a life course in which ambition plays the lead part. In the pages that follow, as in the pages of people's lives, we see how the careers of people chasing dreams—some small, many grand—play out in real time.

* * *

Like the complex careers of the people described herein, this study has followed a path of its own. It is a very great pleasure at the end of a long project to thank the many people who have helped so much along the way. For many years Charles Bidwell, Robert Sampson, and Gerald Suttles have provided invaluable suggestions and support, freely and graciously given. May this work and what follows from it be my thanks to them all.

At the work's inception and at several periods during its gestation, I benefited from people who took the time to talk, write, and listen. With gratitude, I thank Andrew Abbott, Julie Brines, Terry Clark, Robert Dreeben, Wendy Griswold, Donald Levine, Dan Lortie, and the late Edward Shils. Our times did not coincide, but posthumous recognition is due also to some whose writings and social-scientific imaginations helped me to confirm an early interest: W. I. Thomas, Everett Hughes, Joseph Ben-David.

Early ideas and discussions on the subject of this book were presented to groups whose members I thank for their reception and constructive commentary: The Center for Social Organization Studies, the Ogburn-Stouffer Center Education Studies Group, the Culture and Society Workshop, and the Organizations Workshop at the University of Chicago; and panels at annual meetings of the American Sociological Association in Los Angeles and New York. Christine Cassagnau and Patrick Mulvey of the American Institute of Physics skillfully supplied crucial pieces of information. An interim period that I spent with the American Council on Education in 1991 and my discussion with Elaine El-Khawas and Charles Andersen helped in the development of ideas contained in this work.

This study was funded by a generous grant from the National Science Foundation. I thank the foundation for its kind support, and I single out its three anonymous reviewers and Mildred Schwartz for their extremely useful suggestions and the faith they registered in this work.

Doug Mitchell, Senior Editor at the University of Chicago Press, is all that an author could wish for. A supporter throughout, he cheered this work on and made many astute recommendations. In a similar vein I acknowledge the two anonymous reviewers for the press, both of whom served beyond the call of duty. And just when nearly everything about this project seemed complete and the long line of talented people involved with the book seemed at its end, I met with the great good fortune of working with D. Teddy Diggs. Her professionalism as an editor has helped me grow as a writer.

There was a time when I had little sight of this work on the horizon. But then as now, numerous people enriched my life—and I have never forgotten them. Even if the road has been circuitous, the present must give thanks for gifts of the distant past. Thus I warmly sing the praises of teachers from an earlier day: Yvonne Duiker (piano), Steven Gentry (social studies), Joyce Harvey (algebra and geometry), the late Annalee Henderson (trigonometry and calculus), James Langton (music), Anne Riley (English), Barry Rossman (algebra), Emily Schmalstieg (music), Richard Victor (music), and Ron Wilkerson (world history).

Finally, I have dedicated this book to my father, the full measure of whose love, courage, generosity, and companionship I will perhaps never know.

Introduction

Professions are unlike all other groups. They are remarkable for the magnitude of time and toil that people voluntarily invest preparing for and performing in them. The investments that people make toward their professions may be in part economic but are surely also personal. Popular expressions portray people in the grip of work. Whether physicians, attorneys, musicians, or executives, people can be "wrapped-up in" and even "married to" their work. At other times their labor may be described more charitably as a passion, cause, or calling. But whether depicted as an obsession or a love, the work of professionals is typically cast as an endeavor of great commitment, not to mention expertise, and as something from which people derive immense personal meaning and a sense of order.

This book is about how scientists make meaning of their careers. It explores how self-identities of scientists form and unfold in the environments in which they work. In speaking of "self-identity," I am referring to the inner side of an individual, the side shaped by the outside world but normally known by the individual alone. If the private sides of people *are* known by others, they are known only by a small number of confidants whose association is held dear and in deep trust.

Self-identities are distinct from *social* identities, though both are crafted by forces in the environments in which people interact. Social identities refer to how *others* know an individual: as smart, attractive, poor; as a firefighter, a waitress, a student, a poet. By contrast, self-identities pertain to how people know *themselves*. Self-identities refer to people's deepest thoughts, feelings, and wishes. People's self-identities normally differ from their social identities because outsiders are not normally granted access to the parts that make up their "inner sides." Exceptions are of course those outsiders whose access comes by way of a privileged position: close friends, spouses, psychiatrists, clergy. Yet even then people protect the ways they see themselves.

The *professional* self-identities of scientists, as the term suggests, arise and change in specific social arenas of science where practitioners develop conceptions about the symbolic meanings of their work. Professional self-identities are dynamic. They are often infused with gripping sequences of action (rises and falls) and powerful emotions (triumph and defeat), all of which possess social meaning about how careers are experienced. Professional self-identities portray people within the work community in which they have made personal investments and from which they derive meaning about how they see themselves in the big picture of life.

The careers of scientists, like the careers of others, are often marked by poignant drama. In and out of science, careers may be seen as a continuous series of tests and trials, of successes and losses, of hope and uncertainty. A search for coherence is rendered essential for people whose careers are marked by situations, events, and turning points that have led them down twisting paths with felt costs and rewards. People are drawn to compose a coherent narrative (if only for themselves) that sensibly accounts for how their personal biographies have evolved.

Like doctors, lawyers, or artists, the scientists whose career will be described in these pages differ in various ways. They are different ages, and they work in different types of settings, in this case, universities. They have encountered different experiences. In many cases, these experiences have led careers on unexpected routes. Some scientists painfully tell of careers that have foundered. Others recall careers that have scaled the summits of human achievement. In all cases, we will be journeying alongside people who, as firsthand witnesses to their unfolding careers, have arrived at some self-understanding about the ceaseless question, Who am I?

The similarities and differences in how people see themselves bring self-identities into view. Examples of real-life careers illuminate the points where people's paths come together and where they diverge. Although this work reports on a study of scientists, specifically academic physicists, the formation of self-identity and its unfolding clearly pertain to people who earn their living in other ways. Many of the topics that I will discuss and conclusions that I will draw are germane to nearly all arts and sciences in contemporary life: medicine, law, architecture, politics, music, sports.

I will later explain the reasoning that underlies why I have chosen to look closely at physicists. In the process of doing so, I will also explain how this study speaks more generally about many other occupational groups, from architects to athletes. Although physicists are our main subjects, the dynamics of their careers reveal much about all people whose callings similarly involve a search to make significant contributions in the wide world of life. Physicists, like many others, want "to have an impact," to use that overused line. What we have here is not only a close look at what life is like as a physicist but a general social story about seekers who seek to change the world. All embrace *creative rebellion*—an ambition for greatness—which is widely valued throughout our culture, for they have set on an honorable way to influence and advance civilization. We will meet these seekers, hear of their hopes, dreams, and fears, and learn what happens as their searches play out in real time.

Careers in Context: Theoretic Foundations

The internal differentiation of professions merits our empirical and theoretic attention. The environment in which people work, whether a streetside law office or a corporate law firm, whether a community health clinic or a university medical center, sets conditions for a range of factors that effect an individual's social, economic, and psychological well-being. In each of these types of places, members of a profession are exposed to different opportunities for interaction, collaboration, and development, for financial reward and growth, and for self-esteem, intellectual and clinical challenge, and personal happiness.

Traits that establish control over work have long been a focus of efforts by sociologists of professions to delineate contrasts among and within occupational groups.[1] Such traits, which apply to occupations in degrees, typically include restricted access, protracted training, certification, esoteric knowledge, and self-regulation. Some studies have examined how permutations in these characteristics stratify professional communities (for example, Heinz and Laumann 1982, in the case of law; Freidson 1970, esp. chapters 3–9, in the case of medicine). Professions commonly consist of elite and subordinate groups, which differ in their access to prestige, power, deference, and other scarce social rewards.

I am proposing that we think about the internal differences of professions in a new way. This way consists of how the varying social contexts of a profession—here, the different types of universities in which scientists work—set parameters on the ways in which the professional life course is subjectively experienced.

The life course may be understood to consist of the "sequence of culturally defined age-graded roles and social transitions that are enacted over time" (Caspi, Elder, and Herbener 1990, 15). It encompasses "pathways through the age-differentiated life span," where age differentiation "is manifested in expectations and options that impinge on decision processes and the course of events that give shape to life stages, transitions, and turning points" (Elder 1985, 17). In the more restricted usage, "professional life course" focuses on passage in the domain of work and considers the ways in which careers are structured and experienced by individuals and the larger groups to which they belong.[2]

In adopting a life course approach, I will emphasize two key dimensions of careers: time and place. We are going to examine how professional self-identities uniquely emerge in different types of places and how these identities unfold over time. Our concerns are grounded by how scientists *experience* the objective realities of work (Arthur and Lawrence 1984; Bailyn 1989; Berger and Luckmann 1966; Collin and Young 1986; Stebbins 1970; Young and Collin 1992). To that end, we will interpret how scientists view, and have viewed, their unfolding careers.

Self-identities arise from an internal dialogue between what the classic social psychologist George Herbert Mead called the "I" and the "Me" (Mead 1934). The I and Me refer simply to the ability unique to human beings to be both subjects and objects to themselves, capable of reflection. Accordingly, we will focus on how scientists, having been given the occasion to view their careers as objects, account for their careers. *Narrative*, rather than traits, becomes the primary way by which to differentiate among professionals.

In its focus on how individuals account for their careers, this work is concerned with the meanings that scientists attach to their status passages, which of course also mark other occupations in varying degrees. I will be less concerned with the actual practice of science, that is, the work or the technical aspects of physicists' jobs. Several excellent descriptions of the practice of science exist, including Galison (1997), Kevles (1971), Knorr-Cetina (1995), Krieger (1992), Lynch (1993), Pickering (1984, 1995), and Traweek (1988). The work of scientists may even be viewed through popularized albeit fictional accounts, as in Segal (1995) and Taubes (1986).

Ambition

The main theoretic focus of this study is the cultural construct that we know and call *ambition*. As the term is often used throughout culture and as I will use it, ambition refers to a strong will to accomplish. This book thus asks what drives people in their careers, the prime arenas in which accomplishments are sought. It examines how ambition might vary by the places in which people work and by their stages in life. Broadly, this study seeks to reveal what role ambition plays in people's lives, the ways in which ambition is expressed, where ambition comes from, and how ambition might more easily flourish in some places while being constrained, sometimes stymied, in others.

As a concept, ambition has received little systematic treatment and has consequently had a stunted theoretic development. This void may be partly due to an apparent ambiguity in the term itself. Often this ambiguity reflects an uncertainty about whether ambition is "bad" or "good," indicative of ill intent or of a robust life, including a life lived for public welfare and the enlightenment of humankind. Ambition conjures opposing images. On the one hand, the ambitious are praised for their purpose and achievement. On the other, they may be seen as needing to be "ridiculed, taught hard lessons, or brought down with a thump" (Epstein 1980, 7).

Even Aristotle was so mystified by the concept of a will to achieve that he could not resolve what to call it.[3] In fact he used "ambition" and "lack of ambition" to refer to the extremes of the very characteristic that he claimed was desired. A healthy striving mediated by the polar excesses of grandiosity and indolence is what Aristotle ultimately called the "nameless virtue."[4] All the other virtues that Aristotle identified have fast-and-easy labels and convey their meaning in a flash: generosity, gentleness, friendliness, truthfulness, wittiness. But it is ambition—that ever-so-suspect will to achieve—that proves the most slippery both to talk about and to possess in social life.

Gilbert Brim (1992) incorporates a broad definition of ambition in his popularization of how people manage success and failure. He defines ambition as "the basic human drive for growth and mastery" (Brim 1992, 17). The major difficulties in Brim's usage lie, however, not in the breadth of his application but in his assertions that a drive for "growth and mastery" is innate. That is, Brim's account is biologically deterministic in its claims that all people situate themselves against thresholds of "manageable difficulty" reflecting their particular history of success and failure. Brim's stress on individuals as "free agents" comes at the expense of locating them in specific social contexts that shape, sustain, standardize, invigorate, transform, and kill ambition.

John Clausen's (1993) longitudinal study of sixty men and women raises the idea of "planful competence" as instrumental in the quality of life outcomes. Clausen was a key figure in propagating a "constructionist" perspective of the life course: people shape their lives through the choices they make at key junctures and through the overall active orientation they bring to living out their lives. According to Clausen, planful competence comprises three dimensions: self-confidence, dependability, and intellectual investment. Clausen's central argument is that planful competence in late adolescence is predictive of high levels of later attainment and life satisfaction. "The planfully competent adolescent is equipped with an ability to evaluate accurately personal efforts as well as the intentions and responses of others, with an informed knowledge of self, others, and options, and with the self-discipline to pursue chosen goals" (Elder 1993).

Planful competence is not identical to ambition. In principle, a planfully competent person can strive for modest reaches. Clausen, however, found that planfully competent adolescents attain high levels of educational and occupational success in later life compared with other agemates (Clausen 1993, 279, 519–520). Conversely, the ambitious are not necessarily planfully competent. Popular expressions identify such people: those with "blind" or "unrealistic" ambition. We may say, though, that those who *realize* their ambitions are planfully competent: they realize their ambitions "straight out" or modify their ambitions in realizable ways. They possess, along with the necessary material support, the cognitive capacities to persevere as well as to change when change is necessary. Ambition, then, speaks of those individuals who are planfully competent in higher or lower degrees.

The importance of Clausen's work to the study of ambition is therefore indirectly rooted through his notion of planful competence. However instrumental to the quality of adult life, planful competence raises the idea of people being aware of their own intentions and actions. This awareness occurs, perhaps, not at a level where people are capable of evaluating the costs and benefits of all their actions or even of knowing the consequences of "choosing" certain life paths. But an awareness of self—at precisely a level of self-consciousness—plays a critical role in the ability of a person to form visions of the self over time, visions of who one has been, of who one is, and of who one wants to become. Planfully competent individuals are by definition purposive in their orientation to life. Their own life courses acquire meaning by the paths they follow, which are in part determined by deliberate planning, investment, and decision-making. If one is concerned with ambition, then one is also concerned with how people go about constructing, viewing, and pursuing planned (if sometimes wholly unrealizable) futures.

In spite of Clausen's emphasis on life course dynamics, the ways in which planful competence itself is subject to the forces of time and place are difficult to ascertain. Indeed Clausen asserts that planful competence is reinforced when people are launched on a right track in early life. "Launched on the right trajectory, the person is likely to accumulate successes that strengthen the effectiveness of his orientation toward the world while at the same time he acquires the knowledge and skills that make his success more probable. Early advantages become cumulative advantages; early behaviors that are self-defeating lead to cumulative disadvantages" (Clausen 1993, 521).

Unifying the major segments of life—adolescence and adulthood—through a concept unbounded by time is praiseworthy but neglects to take full stock of individual continuity and change (Brim and Kagan 1980; Kagan 1980; Caspi and Bem 1990; George 1993). Much like one might expect of planful competence, ambition need not be viewed as an inherently stable characteristic, for why would one postulate invariance in an individual's drive, stamina, or dreams? Ambitions, of course, change, emerging and growing in one phase while dying or being replaced or reincarnated in another.

Thus in what little attention has been paid to ambition, either directly or indirectly, it remains a creature out of control. This inattention is especially remarkable for societies that attach high value to success, variously defined. In theory, ambition seems to take shape as a function of the times of which people are a part and the places through which they move, but this remains poorly understood. This study deals with the conundrum by posing time and place as two dimensions in which ambition plays or fails to play a part in how self-identity unfolds.

My approach to ambition therefore refutes what many would uncritically take to be conventional wisdom, and readers should be aware of how I use the concept. I do not treat ambition merely as an individual characteristic, unbounded by the contexts of which people are an interactive part. Rather, in addition to being an individual attribute, ambition is socially situated. It is relational, a situational state that is in flux all along the life course and that operates as a function of environment. Contexts of social interaction—the times and places in which scientists work—will be studied to ascertain the ways those contexts establish definitions for individuals and for what they seek to attain.

Careers in Time

The intellectual origin of this study lies in the "Chicago School" tradition of work ethnography inspired by Everett Hughes (Hughes 1958, [1971] 1993, 1994). Hughes conceived of careers in terms of their two-sidedness.

One side consists of the objective statuses a person holds, whereas the other encompasses the shifting, subjective personal perspectives of the career (Hughes 1958). Career turning points operate as mechanisms through which people continually revise their pasts to achieve a coherent and believable sense of self-identity. Turning points entail change not simply in objective status but also in the interpretations that people make of themselves as they move through time.

Drawing on the work of Hughes, I take the theoretic view that self-identity is a product of both past experiences and future expectations. People derive self-understanding from a continuous process of revising and reinterpreting their pasts and their hoped-for futures. People in the present thus represent a mediation of past and future selves (Mead 1932, 1934). "Who will see," Saint Augustine asks in Book 11 of the *Confessions*, "that all past time is driven back by the future, that all the future is consequent on the past, and all the past and future are created and take their course from that which is ever present?"[5]

The scientists in this study have been asked, in effect, to "step outside" themselves. They have provided accounts of who and where they are, where they have been, and where they see themselves headed. In stressing the temporal foundations of self-identity (Wells and Stryker 1988), this work attempts to move beyond a static portrayal that depicts the individual frozen in time (for example, Eiduson 1962; Roe 1952; Strauss and Rainwater 1962).

Studies of the life course have offered theoretic frameworks in which to study the formation and unfolding of self-identity.[6] Of particular importance is Daniel J. Levinson's (1978) major study, *The Seasons of a Man's Life*. Levinson's work established the idea of a "dream," akin in many ways to ambition, as a vehicle for understanding how people organize their passage through life. Levinson drew on the life histories of forty men, between the ages of thirty-five and forty-five, who were employed as business executives, university biologists, novelists, and hourly workers in industry (see also the companion volume, *The Seasons of a Woman's Life* [Levinson 1996]). Levinson put forward a stage theory of adult development in which aging could be viewed as a process that involved periods of "structure building" and "structure changing." In this perspective, the life course consists of a series of alternating stable and transitional periods that shape adult development and conceptions of self.

According to Levinson, a life "dream" emerges between the ages of about seventeen and thirty and represents the key concept around which lives are organized and from which individuals derive a sense of evolving self-identity. The dream, most often rooted in one's occupational life, has the quality of a vision, an imagined possibility of self that generates excitement and vitality. In young adulthood the dream may be poorly articulated and only tenuously connected to reality. For example, it may take a dramatic form as in the myth of hero: "a great artist, business tycoon, athletic or intellectual superstar performing magnificent feats and receiving special honors" (Levinson 1978, 91). In early career stages, people face the developmental task of giving the dream greater definition and finding ways to live it out.

Mentor and personal relationships—presumably the best of such relationships—facilitate progress toward the dream (Levinson 1978, 98, 109). Mentors may act as teachers, sponsors, guides, or exemplars and may provide counsel and moral support in times of stress and self-doubt. They may share in the dream, believe in it, and give it their blessing.

Levinson claims that a midlife transition between the ages of forty and forty-five, popularly referred to as the "midlife crisis," consists of significantly modifying the dream so that it comes closer in line with one's capabilities. Although the dream might embody a sense of omnipotence and excitement, these inspirational qualities also contribute to its "tyranny" (Levinson 1978, 248). Aspects of the dream are dramatized to the point where people find themselves in the grip of its myth. The task at midlife, Levinson contends, is not to abandon the dream altogether but to reduce its excessive power, to make its demands less absolute. Success is thus rendered less essential and failure less disastrous.

The theoretic utility of Levinson's work consists in conceptualizing multiple facets and functions of the self. The dream occupies what Morris Rosenberg (1979) has called the "desired self," or how people want to see themselves in the present and the future. This is set apart from the "extant self," or how people realistically see themselves in everyday life. These sides are in turn distinct from the "presented self," or how people share themselves with others.

"Desired selves" represent idealized self-identities. They encompass status attainment in a broad sense, such as winning a Nobel Prize, designing a world-renowned building, arranging a masterful composition, or winning the respect and admiration of one's peers, friends, or family. In pointing to attainment, desired selves embody those personal ideals that people often keep private (or share only with a select few) and hold as sacred to themselves, for to make one's ambitions public renders them profane and inconsistent with the ideal that cultures place on humility (Epstein 1980).

Whether we speak of "dreams" in Levinson's terms or of "desired selves" in Rosenberg's, the concept invokes idealized self-identities, which play several roles in acting out careers: idealizations may serve as a guide for action and conduct, as a standard by which to work, as a means of tension release, or as an object from which to derive continual motivation, even though people may realize that the imagined identities may be eternally out of their reach. Idealized self-images, in their various forms, serve as potent points of reference that set aspiring individuals on paths to greatness. The image and the path may only come to be a fanciful fiction. But to recall the guiding theorem of a masterful forebear, W. I. Thomas, if voices and visions are taken as real, even as "real fiction," then they are real in their consequences for individuals and social life alike.

People's internal dialogues and their dream worlds change over the course of careers. Following Hughes's emphasis on turning points, the self is susceptible to change as people encounter new roles, situations, and career transitions (Demo 1992; Wells and Stryker 1988). People come to know themselves in different ways, bringing new perspectives to their lives in time and place.

In listening to people's "conversations with themselves," we gain privileged access to how they think, and have thought, about themselves and the world through which they have tried to journey. We enter their heads, as it were, and see infused images about how they envision (or have envisioned) their lives playing out. We hear internal struggles being waged, not all of them easily or quickly resolved, and compromises struck, not all of them in accord with what may have been planned or hoped for. Most central, we enter an internal world, shaped by the outside—the world we collectively know and call our dreams and aspirations. By taking a close look at people's dreams and the self-identities those dreams conjure, we will gain leverage on how people create and act on ambition, as well as what importance they draw from it.

As provocative and useful as Levinson's perspective is for our present concerns, it also bears significant shortcomings. Both Levinson and Rosenberg are reticent about the role that "former selves" play in creating self-identities. In other words, their approaches are *ahistorical*; they do not consider how past experiences shape individuals. Like other "stage theories" of aging (for example, Erikson 1950), Levinson's approach to adult development takes a "cohort-centric" view of socialization (Riley, Foner, and Waring 1988). People are studied outside of the socially contingent circumstances that uniquely shape and characterize them and their life courses (for example, Super 1957).

The ways in which self-identities form and unfold are, however, subject to events and experiences that arise in unique periods of time and that vary from individual to individual. Scientists whose careers span different scientific periods likely account for their careers in different ways, in part because they have encountered different experiences and in part because their socialization is uniquely marked by historical changes that leave an enduring imprint. For example, scientists who enter a field at a time when jobs are plentiful might well account for their careers differently than do those who enter when jobs are scarce. Shared events and experiences lead to the idea of a cohort or generation whose members are characterized by defining moments or periods that morally set them apart from those who have passed before and from those who are yet to pass (Mannheim 1952; Ryder 1965).[7]

Elder's studies on the Great Depression have raised the importance of inter- and intracohort comparisons (Elder 1974, 1981). His work has addressed the linkages between social change and life paths and how these paths have in turn affected psychosocial functioning. The depression was not experienced uniformly by different birth cohorts. The timing of the depression relative to developmental age placed one cohort that came of age during the depression (1928–29) at a greater risk of impaired life chances and development than a second cohort whose members were older by seven to eight years.

In describing the careers of scientists, I will also use cohorts, which will allow for a more complete account of how people's self-identities arise in light of varying social experiences related to age and maturation. Specifically, I will use cohorts in two ways. First, I will use cohorts to draw distinctions in how the young, old, and middle-aged craft self-identities and perspectives on their careers, perspectives that differ systematically from one another. This use of cohorts draws attention to the sociocultural meanings that people assign to their careers given their age and past experience (Neugarten 1979a; Neugarten and Datan 1973; Wells and Stryker 1988). Second, I will use cohorts to selectively examine the impact of macrolevel social change on individuals. This use of cohorts is similar to studies, such as Elder's described above, that examine how life courses take shape as a function of the historical times of which people are a part.

In using cohorts, I hope to avoid depicting the science community as a homogeneous mass and to avoid casting the science career as a sequence of monolithic stages through which all scientists invariably pass. "Dreams" take different forms; for some people, they may scarcely exist. But where they do exist, dreams play out differently for different people as a result of the environment and the time. Simply put, scientists age in different ways; their careers develop or fail to develop at different rates and as a result of different circumstances. The incorporation of cohorts into the design of this study permits a comparison of how individuals within and across age strata "make meaning" of their careers in various ways.

Careers in Place

Studies of the life course have been less than informative about how self-identities arise in specific contexts of work. Levinson used life histories of people from occupational groups, but the people were studied as if they were detached from these groups. Studies of adult development have sought to derive *universally* experienced stages, defined by periods of innate challenge or "crisis" (for example, see Baltes 1979; Erikson 1950, 1959, 1982; Levinson 1978, 1996; Sheehy 1976). They have been less attentive to how self-identity emerges out of the particular settings in which people work.

In an important article, Dale Dannefer (1984a) sparked debate by proposing that the study of adult development was in need of theoretic reformulation.[8] Dannefer argued that the prevailing mode of studying the life course—the *ontogenetic* model—"is not an appropriate foundation for its subject matter because it tends to treat the individual as a self-contained entity and fails to recognize the profoundly interactive nature of self-society relations and the complexity and variability of social environments" (Dannefer 1984a, 100). One must indeed question a mode of inquiry that leads to such invariance in adult development and socialization, a mode most readily apparent in stage theories of aging (Erikson 1950; Levinson 1978, 1996).

The alternative is a *sociogenic* approach (Dannefer 1984a), which grounds the study of adult development in the highly variable and contingent social contexts of which people are a part.[9] By placing people in their social contexts, we elevate the study of development by addressing how development may be *different* among people as a result of their exposure to various opportunities, constraints, and turning points.

We are led, therefore, to the social contexts in which science is done, in which scientific lives are led. The literature in the sociology of science is rich and diverse.[10] Studies have addressed comparative failure in science (Glaser 1964a), the backgrounds and careers of scientific elites (Zuckerman 1977), the centrality of originality and competition in science (Gaston 1970), the impact of rapid discovery on the scientist's career (Reif and Strauss 1965), and the consequences of specialization and specialty change in the scientific field (Ziman 1987). Other studies have depicted the science community in broader terms, their special themes encompassed by a view of science as a social system (for example, Barber 1952; Glaser 1964b; Hagstrom 1965; Hirsch 1968; Storer 1966).

A heavy stream of research that has examined the nature of stratification in science is most germane to our present purposes because these studies, like this book, examine different "strata," or contexts, of science (Allison, Long, and Krauze 1982; Allison and Stewart 1974; Cole 1970; Cole and Cole 1967; Crane 1965, 1969, 1970; Fox 1985; Gaston 1978; Hagstrom 1971; Hargens 1969; Hargens and Hagstrom 1967; Long 1978; Long, Allison, and McGinnis 1979; Long and McGinnis 1981; Reskin 1977, 1979; Zuckerman 1970, 1977). Jonathan Cole and Stephen Cole's (1973) *Social Stratification in Science* is perhaps the most comprehensive treatment within this body of work.

Much of the research on stratification has been inspired by Robert Merton's theory, developed over fifty years ago, about the normative structure of science. According to this theory, science has an "ethos" consisting of the norms of universalism, communism, disinterestedness, and organized skepticism.[11] The outpouring of stratification research has stemmed from a sustained interest in the norm of universalism, which stipulates two related requirements. First, when a scientist contributes to knowledge (primarily through publication), the science community's assessment of the merits of the contribution should not be influenced by personal or social attributes of the contributor. Second, a scientist should be rewarded in ways that are commensurate with the measure of a contribution. Universalism is contrasted with particularism, which involves the use of characteristics, such as sex, race, age, or political party affiliation, as a basis for assessing contributions and rewards.[12]

No one will dispute that there is great inequality in science, as in other social institutions. Researchers of stratification have been concerned more specifically with whether this inequality results from the fair use of universalistic criteria or from the functionally irrelevant use of particularistic criteria. That is, the research has addressed whether inequality in science is equitable (just) or inequitable (unjust). Along similar lines, more recent efforts have taken a more sophisticated tact and asked when, and under what conditions, particularistic criteria are invoked (when, ordinarily, universalistic criteria would apply) to form opinions and make decisions about people and their contributions, usually with significant consequences for person and career alike (S. Cole 1992; Long and Fox 1995). Thus, researchers have investigated precisely how the reward system in science operates and what consequences its functioning has on matters such as the job placement of Ph.D. graduates, promotion, tenure, productivity, recognition, and other foci of participation and attainment in science.

As much as these studies have told us about the process of stratification in science, we know remarkably little about the details of "what life is like" on the inside. Most of the stratification research has been undertaken with cross-sectional surveys, which, though possessing the virtue of breadth, fail to provide a picture of the finer-grained cultural dynamics at play in any given setting where science is done. The stratification research has been almost exclusively quantitative, often seeking statistical models to predict measurable outcomes in careers, outcomes such as publication productivity and citation rates. In addition this literature, and the larger family to which it belongs, have largely bypassed issues of "self" and identity.[13]

The people behind the science have largely been forgotten. But it is they who attract wide interest. Not since Anne Roe's study in 1952, *The Making of a Scientist,* and Bernice Eiduson's 1962 work, *Scientists: Their Psychological Worlds,* have we become acquainted with what scientists' lives are like. Yet those works were heavily oriented to psychology. They sought to render characterizations of the personalities of scientists through various psychometric evaluations, including the Rorschach (or "inkblot") tests that were used to elicit mental associations with abstract images. Furthermore, these works said little about the life course, which has since become a leading theoretic basis for examining lives in context (Bronfenbrenner 1979; Elder 1992; Moen, Elder, and Luscher 1995). *The Stars Are Not Enough* "brings people back in," and does so with explicit concern for the ways in which their careers acquire meaning in time and place.

Although the "ethos" of science may speak about the norms and values that guide *all* scientists in their conduct toward research and researchers, whether or not a scientist is active in research, the "ethos" does not speak about the norms and values that distinguish various settings of science. In one setting, a drive to succeed in research (to pick just one activity in science) may be the norm, cherished and loved by nearly all, whereas in another setting, research may be strongly discouraged and may cause raised eyebrows.

Studies of stratification in science have typically approached their topics with a mind-set that underscores a set of norms and values most common among elites. This tendency has likely resulted from the vehicle that most have used to enter the stratification fray: the norm of universalism, which emphasizes *research* productivity. Yet a vast portion of the science community is not active in research (Gustin 1973; Zuckerman 1988); thus, explaining performance solely on the basis of how effectively universalism operates in science is inappropriate. Scientists, like people of other social institutions, march to many different drummers, of which only one sounds the continuous beat of research expectation. To extend the metaphor, the drummers to which scientists march likely change over time because of factors that universalism does not govern: that is, people's work habits, interests, motivations, and ambitions change as their careers (and personal lives) unfold. Science may be universal in the *scruples* it prescribes for its practitioners, but its practitioners are highly particular in the kinds of orientations that they develop toward their work.

There is thus a lingering sense in the stratification literature—if only because of the very use of the term "stratification"—of winners and losers: those beneath the head of the totem pole are seen as somehow not quite as good as those at the top. One should pause to consider, as we shall here, not only ways in which the vast majority of scientists may not in fact perform like elites but also ways in which they may perform far better. In the end, of course, it is left for individuals, not sociologists of science, to render such judgments. The point remains that science is internally differentiated along several dimensions, not just research productivity, and in contrast to many extant sociological studies, this study will examine science in ways that take these several dimensions into account.

I too will investigate different contexts or strata of science, but my focus will be on the ways those contexts shape the narratives that people tell about their careers. I will speak of "settings" and "contexts" and "regions" rather than "strata," which imply a pecking order that obscures the meaning that people themselves use to organize their career narratives, imposing a set of meanings from without.

Throughout the discussion, we will see that contexts are often distinct, in ways that the term "strata" (as in separate levels) accurately captures. But we will also find that contexts are sometimes similar, bleeding over into another. In these important respects, "strata" becomes an inappropriate term for thinking and speaking about the similarities and differences in academic environments. Instead, for reasons I will explain in the next couple of chapters, we will explore and refer to "social worlds" of science.

To put this conceptual matter aside for the moment, this study may be read in terms of the interplay between stratification, rewards, career imagery, and construction of self. We will "go into" different strata, seeing the ways in which they distinctively shape conceptions of self and career (cf. Kohn and Schooler 1983). Issues that the stratification literature has dealt with statistically—productivity, reward, location, participation, and the like—will all be prime ingredients, but they will surface primarily as factors in the ways that people variously narrate their careers and create self-identities. Thus we will come to know strata in finer detail, acquiring a sense of how careers unfold at different locations in the spectrum of science.

Echoes of Time and Place

By taking hold of what *achievement* means for different people, we can depict more clearly how people experience and account for their careers. Clearly, not all scientists are motivated by the same forces. Ambition, for example, is not uniformly manifest in all scientists.

Ability partly accounts for differences in the need to achieve, but this need is also likely conditioned by the social contexts in which scientists work. Conceptions of achievement, of extraordinary achievement, and of lack of achievement are derived from where scientists work in the institution of science, which in turn determines the audiences and reference groups that scientists use to form images of themselves and to measure their self-worth.

Merton first conceived the idea of "locals" and "cosmopolitans" in his study of community influence (Merton 1957). Gouldner picked up on the distinction to characterize members of organizations, including faculty members in universities (Gouldner 1957–58). Local and cosmopolitan identities are, according to Gouldner, based on three criteria: loyalty to organization; commitment to specialized skills; and orientation to reference groups. *Locals* are people with high loyalty to their employing organization, low commitment to specialized skills, and internal reference groups. By contrast, *cosmopolitans* have little loyalty to their employing organization, high commitment to specialized skills, and predominantly external reference groups.

Though provocative in its analytical distinction between the types of orientations that people have toward their work, this dichotomy is oversimplified. The distinction is less powerful when attempting to characterize a profession that is internally differentiated along numerous dimensions. Personal experience, career turning points, and historical influences on cohorts stand among these unaccounted-for dimensions.

But even more pressing is the atemporality of the local/cosmopolitan distinction. The distinction separates individuals into camps (which have in their own ways come to possess derogatory connotations) but fails to consider individual change over time. *The Stars Are Not Enough,* by contrast, places identity in a dynamic, rather than a static, context. Each dimension supporting the local/cosmopolitan dichotomy is fluid: orientation to work evolves in response to career turning points, different environments, the reevaluation of priorities, the attainment of certain levels of success, and the like. The formation and unfolding of self-identity is therefore more complex than the local/cosmopolitan characterization depicts.

Finally, we can draw some parallels between the social construction of scientific knowledge and this book. The social construction of scientific knowledge deals with how "truths" are made through selective social processes and conditions. It is a tradition that questions any sense of "objectivity" in what we call scientific knowledge. Instead it emphasizes the subjective elements involved in the production of what the science community certifies as valid and true. At the extreme, social constructionists argue that there is no cognitive basis of knowledge, that knowledge has no external reality. This study, however, examines the construction of *careers* rather than knowledge, and so my discussion of matters that flow from the latter vein will not satisfy an appetite in search of a full-course meal. Nevertheless, despite differences, there are also some connections between this study and those in the constructionist tradition.[14]

The greatest common ground lies in what I have posited to be the two social dimensions along which careers are shaped and experienced: time and place. Just as I will address the ways in which career narratives take a form as a function of the times and places through which careers unfold, social constructionists have addressed how time and place have selectively shaped what is called "knowledge." Thus, whether we are talking about careers or knowledge, time and place are seen and studied as shaping influences.

For example, Steven Shapin (1996) has investigated the "authenticity" of what has been called the Scientific Revolution. Instead of being a discrete event that would have warranted the use of the phrase, the Scientific Revolution was instead, Shapin argues, "a diverse array of cultural practices aimed at understanding, explaining, and controlling the natural world, each with different characteristics and each experiencing different modes of change" (1996, 3). The term "Scientific Revolution," he claims, was probably coined in 1939—not in the seventeenth century—to summarize, and give legitimacy to, a series of complex and sometimes contradictory practices that over time developed into what came to be called "the scientific method," the so-called objective process of experimentation and deduction that leads to scientific "truth" about the natural world. The Scientific Revolution is thus presented not as an objective truth but as a historical construction.

Many other studies have investigated the shaping influences of time and place on specific knowledge or "facts," underscoring how this knowledge was allegedly constructed by being situated in unique social circumstances (for examples of constructionist work, see Latour and Woolgar 1979; Lynch 1993; Pickering 1995; essays in Jasanoff et al. 1995 and in Pickering 1992). Echoing themes throughout much of the constructionist tradition, the essays in Adelle Clarke and Joan Fujimura (1992) examine how knowledge is resource-dependent: the outcome of any given scientific inquiry is seen to be shaped by constraints imposed by workplaces, theories, research materials, audiences and consumers, and scientists themselves. "Tools," "jobs," and the "rightness" of tools for the jobs are socially constructed within restrictions imposed by time and place. Thus our contemporary understanding of genetics is seen to be contingent on flatworms, which were common "tools" in genetics research in the early twentieth century (Mitman and Fausto-Sterling 1992), and our understanding of antibodies is seen to be contingent on research techniques used to establish germane scientific concepts and facts in the late nineteenth century (Keating, Cambrosio, and Mackenzie 1992).

Thus, according to constructionists, knowledge stripped of time and place has no basis for being; correspondingly, people detached from time and place lack the very "tools" necessary to craft identity. Just as knowledge about the natural world is seen to have a context, so does the career, and it will be through a close look at careers in context that we will come to know scientists as people—how they order and make meaning of the natural world and of their journey through it.

Why Scientists?

A legitimate question at this point is, Why study scientists? If my aim is to understand how professional self-identities emerge and unfold in the contexts in which people work, why not talk to lawyers or to doctors or to some combination of professionals?

Artisans, Routine, Open-enededness

My selection was guided by my theoretic concerns. Science, more than law or medicine, is an artisan profession. The *content* of what scientists do is highly reflective of the scientists themselves. Scientific work is a personalized craft in which the best products of labor are believed by scientists to be enshrined with aesthetic value (see Krieger 1992 and Traweek 1988). The experiment, the piece of equipment, or the idea that embodies beauty, grace, style, and originality reflects the craftsman's well-cultivated "taste" or "judgment," words that scientists frequently use to characterize their work and that of others (McAllister 1996). The notion that scientists consider scientific work to be aesthetically pleasing contradicts popular stereotypes that depict the work of scientists (and the scientists themselves) as being formal and rigid, following lockstep procedures in which the affective elements of researchers fail to enter their labors

(LaFollette 1990). Much to the contrary, words like "taste" and "judgment," and accounts of people who are known to possess and not to possess these qualities, make up a system of meaning that is de rigueur for scientists (Chandrasekhar 1987; McAllister 1996). These aesthetic qualities figure prominently in how scientists select problems and form strategies toward solving them. To know about a piece of scientific work is to know about several people behind the work.

The work of other professions is considerably more routinized than science (Hargens 1975, esp. chapter 3). Law and medicine are both governed by strict bureaucratic codes and standardized operating procedures. If there are venues for personal creativity, they are few. Indeed, individualism often elicits charges of showmanship, incompetence, malpractice, and harassment. By contrast, scientists operate autonomously, free to choose and approach problems at their own discretion, albeit within established conventions of what the science community accepts as valid and true.

Further, careers in science stand apart from many other careers on the basis of a peculiar and sometimes daunting characteristic: their open-endedness. Unlike law or medicine careers, which assume highly standard forms in which the individual practitioner expects fairly regular tasks, science careers are heavily shaped by the *creative* efforts that individuals put into them. Although autonomy in science (and in the academy more generally) has been popularly depicted as a luxury, it is also a curse. No organizational imperatives tell the scientist, unlike the lawyer or the doctor, that he or she has to get up each morning to work. And after each day of effort, the scientist is left not completely knowing what those efforts will bring. The career is led by the eternal hope that the efforts will lead to something, but the exact nature of that something is not known until well after the fact. Prizes in science, including but by no means limited to the Nobel Prize, are customarily awarded decades after the prize-winning work was done. Less momentous but still crucial measures of performance in science are equally fleeting. It is often difficult to gauge the importance of a paper, the everyday staple of many scientists' work, until well after it has been published. Even in teaching, the extent of one's effectiveness and influence can be eternally questioned.

The lawyer wins a case; the athlete wins a match or game; the doctor cures a patient. The day comes to a close, and in each of these instances, the individual knows for certain what was won or lost and knows for certain where, therefore, he or she stands. Before long, another day starts anew.

The open-endedness of science careers makes the future look painfully uncertain. The scientist does not know what, if anything, is coming next, or even when and how the career will end. The self-identities of scientists differ from those of other professionals in the ways in which scientists cope with this uncertainty. Uncertainty, together with its torment, is after all what motivated Max Weber to conclude that one must be *called* to science: "For nothing is worthy of man . . . unless he can pursue it with passionate devotion" (Weber 1946, 135). The distinctly personal relationship between scientists and their work renders science especially fertile ground on which to explore how self-identity inheres in the work that people do.

Knowing through Narrative: Detail and Repetition

In principle, I could have studied scientists from any one field of science or scientists from some combination of fields, but I selected scientists from only one branch: physics. This selection was, again, the result of theoretic considerations. I am placing the formation and unfolding of self-identity in the context of a moral order. Moral orders—what groups define as normal and desirable—differ dramatically among occupations. I am "going into" the worlds in which physicists live. These worlds instruct them about how they should define effort and achievement, about what they should expect of themselves and of others, and about how their careers should look at certain stages. By confining this study to one occupational group, I am "controlling for variation" that may stem from the distinct conventions and sets of expectations peculiar to single occupations.

I have confined my selection of physicists to the academic sector of science in order to maximize depth of analysis. One of the hallmarks of qualitative studies is fine-grained analysis of the categories that people use to establish meaning. With such studies, the greatest understanding accrues through two interlocking means: *detail*, which is used to account for individuals and their social worlds (which usually means drawing on lengthy quotations that place people in context); and *repetition*, in which patterns increasingly come to light as they recur in narratives.

A number of physicists are employed in industrial and government sectors. But including them in this type of study would mean sacrificing the extent to which we will come to know the individuals who make their livings in academic worlds.[15] Moreover, comprehensive studies of nonacademic scientists have already been written (Glaser 1964b; Kornhauser 1962; Marcson 1960). By confining the study to academic physicists, I have purposely aimed at maximizing the variety of careers across different contexts *within* one sector.[16]

In the Shadow of Gods

My justification for studying scientists, particularly physicists, rests on one further point, perhaps the most important of all.[17] In what social scientists and others have called the era of modernity, scientists have assumed a privileged place like that of prophets in religious eras and poets in the classic city-states (see LaFollette 1990; Meyer et al. 1997). They are regarded as having a kind of genius that sets them apart from ordinary mortals and, until proven fallible, exempt from constraints on their range of inquiry.

Their attributed genius is to a large degree created by a feature that in many respects makes physics exemplary, distinct from all other fields and from all other endeavors: the alleged certainty on which the work of physicists is based.[18] Physics is a field, unlike fields in the social sciences and humanities and other fields in the natural sciences, in which its practitioners perceive one another as either right or wrong (see Traweek 1988, 1–5). They are right or wrong in measurement, in calculation, in interpretation. Physics is said to be distinguished, then, by a lack of "gray areas," in which arguments may be framed in matters of degree or on the preponderance of evidence, not to mention the political motivations or subjective biases of an investigator. Arguments in which interpretation rests on the burden of proof are a luxury (or, as others would assert, a limitation) of researchers who cannot control, or ever adequately measure, all aspects of the populations or topics they study.[19]

Especially in the last half century, scientists have replaced prophets and poets as the great minds that we turn to in order to fathom and manage our own life and times, our own pasts, our own futures (Meyer et al. 1997). Vast sums of money are spent on science, much of it basic science, where importance, relevance, and applicability are rarely detectable. Hardly any of this science is comprehensible to those furnishing the money. Yet the dawn of modern times with the seemingly ever-intractable modern problems has drawn society to ask plaintively whether science can save us (Lundberg 1947).[20] Scientists are largely accountable to themselves, and expectations of them are, like expectations of prophets and poets, unlimited. Like the achievements of prophets and poets, then, the accomplishments of scientists are mythified in both popular and historical accounts that impart to scientists a life course in which only Herculean accomplishments mark one's passage.

The open-endedness of a scientific life creates an uncertainty about scientific accomplishment, in turn leading to a fundamental uncertainty of self among scientists. Scientists' uncertainty of self is magnified by the mythification of the scientists who have preceded them. Mythification is especially prominent in physics. Physics is society's science par excellence. It is often taken to be society's oldest science (unseated only by politics if we count Aristotle); it is said to be the most exact, the most mathematical, the most objective, and the hardest of all the hard sciences. Even in the popular mind, physics has a recognizable genealogy of immortals: Copernicus, Einstein, Keppler, Newton, Ptolemy—a secular pantheon.

The mythification of scientists is reinforced through a burgeoning cultural practice: the production of scientific biographies and autobiographies. There are dozens: one on Luis Alvarez, another on Victor Weisskopf; one on Herbert York, another on S. Chandrasekhar; a few on Lewis Thomas, a few more on Freeman Dyson; several on Sheldon Glashow, many more on Richard Feynman.[21] Rarely are such works written about, or by, an obscure scientist. They are written about elites, and then only about elites who are older, and they discuss the subject's life and work from the illustrious position of the conclusion of a superlative career.

These works affirm the heroism involved in carving out a scientific life, and they thus perpetuate the dream of great attainment by providing exemplary cases that sustain that dream (Carlyle [1904] 1966; Schwartz 1985). Yet few biographies help us understand the contexts in which physicists work; rather they point to the heroic self-concept that is created by all who venture into science. All scientists who are the stuff of biography stand on the shoulders of giants, and though they willingly acknowledge the ties that bind them to others, the practice of biography individuates effort, creating individual heroism out of highly collaborative undertakings.

With few exceptions, scientific biographies and autobiographies tell us little about the intimate understanding that scientists come to have about themselves and their careers. Biography makes identity public, and the subject's thoughts and words are filtered with a public audience in mind. Such will not be the case here, where scientists have spoken on the condition of anonymity.

Scientists themselves (along with many sociologists of science) are often frequent consumers of scientific biography and have remarked on the lack of insight that such works shed on what many would take to be "matters of the heart." The great astrophysicist S. Chandrasekhar (whose biography, incidentally, stands as one of the few exceptions to the pattern) has remarked: "I would love to have known what Einstein's attitude was at the end of his life. My own suspicion is that he was not very happy. I have read a great number of biographies. They never give answers to what scientists like Maxwell, Lord Kelvin, Stokes, or Einstein felt" (Walli 1991, 305). Similarly, the sociologist of science, Warren O. Hagstrom, has observed: "Scientists' autobiographies are often unsatisfactory for socio-logical purposes. When scientists write such things, they carry with them the scientific norms of objectivity and detachment; since their social rela-tions are not expected to influence their scientific judgments, they fail to re-port much about them" (Hagstrom 1965, 2–3). Quoting a typical statement from Darwin, Hagstrom underscores his point: "My chief enjoyment and sole employment throughout life has been scientific work; and the excite-ment from such work makes me for the time forget, or drives away, my daily discomfort. I have therefore nothing to record during the rest of my life, except the publication of my several books" (Hagstrom 1965, 3).

"Matters of the heart"—what are they, and more important, what is their larger significance? The expression comes—so far as I am con-cerned—from Joseph Epstein's memorial tribute to Edward Shils, the so-cial analyst of higher learning:

> Edward taught me . . . how to laugh in dark times. . . . Four or so weeks before he died, I told him that he had been a very good friend, the best friend I expected ever to have. He told me that I had been a dear friend to him as well and added the qualification: "even though we rarely spoke together of things of the heart." I tried not to show shock at this, and I have since pondered its meaning. . . . What he meant, I have come to be-lieve, was that we chose not to speak to each other, except in the most fleeting way, of our doubts and disappointments and griefs. We didn't, in part because neither of us was therapeutic in our impulses—better, we both felt, to eat than spill the beans—and in part because we each lived by a code in which a man does not whine or weep, even on the shoulder of his dearest friend. . . . Edward applied his own high stan-dards to himself, and if he found so many contemporaries wanting, he knew that, as good as he was, he was not as good as he wanted to be. He left unfinished vastly ambitious manuscripts; and as excellent as his published work was, he always felt it could be better. . . . Powerful as so much of his writing is, and permanent though I believe his contribution to the study of society has been, he would, I do not doubt, have preferred to have left behind a single masterwork. But, alas, he didn't. It was not something we talked about. He hungered, too, for a richer family life than he had. . . . By the time I knew him, Edward had long been di-vorced, though he had had chief responsibility for raising his only child. . . . Edward told me that he felt he had, over the years, acquired the character to face death without terror. And so he did, little as he wished to leave life. . . . Until the end, he continued to see students. Old friends, my wife among them, came in a week or so before his death to bid him good-bye. As my wife said, he did all the work: telling her how

much her friendship had meant to him, how he would miss her, instructing her not to grieve for him but to remember the lovely times they had had together. I had taken to kissing his forehead when I came into or departed his bedroom. He would lift my hand to his mouth and smile. . . . Any but a fatuous man will be disappointed by his own life—will feel that he could have done more and better than he did. The truer test is whether a man has disappointed others. (Read 12 April 1995 at Bond Chapel, University of Chicago; published as "My Friend Edward" [Epstein 1995, 392–394])

In large degrees, then, myths mold the careers of all people, who can hardly be expected to turn away from the illusions of grandeur that great figures inspire. The fact that revolutions are rare is a minor point; all people, at some time, find themselves tempted by these illusions. Most of those who cease to be tempted have knowingly put these illusions aside. Because demigods are part of the direct professional parentage of those who enter the field of physics, the mythification of careers is especially prominent in this field. If there is any profession where ambition places one directly in the shadow of gods, it is physics. Thus if we want to find out how one lives with this sort of company and how a scientific life plays out in real time, physics is an ideal place to start.

Ambition in Our Time

For all the reasons that motivate this study of physicists, it would be foolhardy to suggest that physicists alone embody ambition for greatness. Although I have put forth the benefits of examining a single occupation in the context of the moral order that guides and shapes its members, and although I have justified the selection of physics on the basis of its kingpin status—the crème de la crème—among the sciences and in the public mind, we should treat that which drives physicists as a more general symptom of the wider culture.

We think thus of artists, architects, physicians, lawyers, composers, poets, politicians, and athletes. All of these occupational groups are set in a class on the basis of select criteria that the wider culture uses as an ultimate judge of how careers in these fields add up: great originality, ingenious invention, and heroic accomplishment. In identifying groups of such diversity, we see how widely diffused the ambition for greatness is as a cultural theme. Few other marks distinguish in quite the same way a people with such majestic longing, what has been called "the achieving society" (McClelland 1961).[22]

Like physicists, each of these groups has its own recognizable pantheon, legendary figures who inspire peak performance through their own exemplary professional passages.[23] For artists there are Van Gogh, Picasso, Renoir; for composers, Bach, Mozart, Rachmaninoff; for athletes, Michael Jordan, Babe Ruth, Walter Payton. They and others like them conjure rousing images and scenarios: the enigmatic self-portrait, the moving musical score, the Olympic-like game-winning play.

The heroism that envelops these pantheons spurs a quest on which any serious newcomer must venture, for to reject the call outright would amount to an altogether second-rate beginning. What could one accomplish in football, music, medicine, or physics without being tempted by grand and mighty voices and visions? Those voices and visions point the way, while at the same time they mercilessly crush people in periods of self-doubt. They embody an ideal, an imagined possibility of what people can become.

Ambition for greatness becomes a narrative unto its own, a steering voice and vision. It speaks a sentence that rouses the passions, telling how the career should go and what one must do to assume an eminent identity. It provides nothing short of a guide for living. Voices and visions are heard and seen in physicists' labs as much as they are on West Texas football fields (Bissinger 1990) and on inner-city basketball courts (Joravsky 1995). Whether they are Nobel or hoop dreams, the voices and visions reveal to us the romantic cultural narrative that is ambition.

This, then, is not only an in-depth look at the careers of physicists but a more general cultural account of the many careers in which ambition for greatness plays the lead part. Though physicists will be our subjects, and we will get to know them in detail, the outlines of their careers tell a broader social story about the many people whose work similarly sets them on paths to greatness. The drama lies in knowing that some will "make it" and others will come close, whereas still others will undoubtedly fail. In what lies ahead, ambition for greatness plays out not merely as an individual phenomenon but as a function of the times of which people are a part and the places through which they move.

Conclusion and Overview

I have argued that professional self-identity must be examined in the context of the places and times in which careers are experienced. Studies of professions have neglected to address internal differences among members, variations that arise in the distinctive worlds from which they derive meaning about how they see themselves. I have proposed to remedy this defect by focusing closely on how self-identities among scientists emerge and unfold in their work contexts.

In chapter 1, I discuss the sources of data and the research methods on which this work is based. I will account for how I have gathered life histories, in addition to other data venues, from scientists whose careers differ by time and place. I will conclude by taking a critical look at the type of data and methods I have used, assessing their strengths and limitations.

In chapter 2, I enter the worlds in which scientists work. I will describe these worlds by highlighting not only their similarities to but also, and even more so, their differences from one another. I will accomplish this by examining careers that are seen as normal by people in these worlds. Delineating similarities and differences among worlds allows us to understand the conditions under which careers are experienced. In looking at each world's customary practices, we learn what life is like on the inside.

In chapter 3, I discuss how scientists account for their pasts. I focus on their early aspirations and on how these aspirations changed from the points at which they started out in science up to where they find themselves today. My intent is to explain how their outlooks have evolved in light of encounters with events and situations that altered their perceived prospects of mobility.

In chapter 4, my focus shifts from scientists' retrospective accounting of their careers to their prospective accounting. I describe the visions they have of their futures in light of the social circumstances in which they find themselves. I will explore the forms and functions of career fantasy in an effort to explain the role that dreams play in the acting out of careers.

In chapter 5, I discuss the career problems that scientists experience. I focus on the issue and role of self-doubt. In particular, I will be concerned with how self-doubt may be unevenly experienced by scientists and with the way in which self-doubt may operate to form and fuel, rather than inhibit, ambition. I will also pay special attention to women's careers in science, exploring how they are similar to and different from men's.

In chapter 6, I conclude my discussion by studying the role of ambition in science careers. I discuss ambition in its sociocultural context by describing the conditions that foster and inhibit it and the consequences that these conditions raise for the unfolding of careers. Finally, in chapter 6, I will also introduce the idea of the master narrative, a device that conceptualizes how people organize and publicly present their life passages. These narratives capture regularities of rhetoric and self-identity flowing from the very regularities marking the environments in which people carve their careers. This device will thus be offered as a sociological tool, open for wide use throughout culture, not only to explain differences between the shaping environments that people call "work" but also to reveal how people understand themselves in light of their formative travel through time and place.

ONE

The Study and Its Context

This work is the product of highly cooperative individuals who, under assurances of anonymity, patiently spent many hours talking to me about some of their most personal feelings and judgments on themselves and their careers. Over a period of about six months, I traveled to six universities across the United States and interviewed sixty scientists about their career experiences. All of the scientists were selected at random from the physics departments in these schools.

In each of the schools, I talked to roughly equal numbers of scientists of three cohorts. Each cohort was established a priori by year of Ph.D. The first cohort consists of scientists who received their Ph.D.'s before 1970; the second cohort comprises scientists who received their Ph.D.'s between 1970 and 1980; and the third cohort includes scientists who received their Ph.D.'s after 1980. These cohorts were intended to cover a period of time in which scientists could speak from points at early, middle, and late career. Moreover, the cohorts allow an assessment of how careers are molded by the sociohistorical conditions that distinctively mark groups of individuals making passages during different intervals of time.

Department chairs in each of the physics departments were sent a letter that described the overall purpose of the project and detailed what a professor's participation in an interview would involve. The purpose of this letter was to inform the chairs of my intended visit, alerting them of my presence and the scope of my activities. Moreover, I wanted to inform department chairs of the legitimacy of the project should faculty members inquire of them about solicitations for interviews. All of the chairs were invited to contact me if they had questions or concerns about the project or my visit, though none did.

Each of the scientists was contacted first by letter, which similarly described the study. About one week later, I contacted the scientists by telephone, reminding them of the letter they had received, describing more fully the study and their possible participation in it, and asking if they would agree to be included. Some agreed to participate without further ado; others wanted to have specific questions answered, and then agreed; others declined.

The overall "rate of response" to the study, that is, the percentage of scientists who agreed to participate in the study after being asked by phone, was relatively high: 70 percent. Those who declined to participate normally did so without reason, possibly not wanting to participate in a study that they realized would call for personal judgments and the reliving, or at least retelling, of consequential personal experiences, including hardships, anxieties, disappointments, and so forth. It seemed inappropriate to ask these individuals why they would not participate, even though such information is of course useful in understanding the ways in which the resulting sample may not be wholly representative. To press the point would in effect have put the individual on the spot for something that was by definition voluntary. Some who declined offered the reason—ubiquitous among professionals—of not having enough time or of being out of town or out of the country.

Yet the relatively high response rate points to the comments made by those scientists eager to participate. Several of the scientists expressed their sensitivity to the problems of doing research and quickly agreed to help with research in which their participation was vital. More scientists seemed, frankly, intrigued by the study itself—a rarity in their own kind of work and an occasion in which to delve into matters of their own intimate concern. The interviews took some time, averaging ninety minutes in length. After completing their interviews, most of the scientists expressed an interest in receiving some follow-up about the project and its results.[1]

The Interviews

I asked the scientists questions about their professional biography and how they have experienced their careers. They provided detailed accounts of how they view themselves in relation to their work. Their narratives are laced with stories of the situations, circumstances, turning points, and events that have shaped their professional self-identities.

The questions I asked were divided into six interrelated sections, each pertaining to social-psychological aspects of career experience. I wanted to describe scientists' careers in the context of the moral orders in which they work. To do so, I asked the scientists about what it means to "succeed" and what it means to "fail." Questions about relative success allowed the scientists to define and characterize what an "average" or "normal" career looks like for the people they work with. Likewise, they told me how careers vary from the average and noted where they saw themselves standing with respect to this variety.

I asked about "career fantasy," what scientists dream about professionally. This was done to elicit images of how they see themselves at future points in their careers. In the course of discussing such images, the scientists disclosed their aspirations. They told of the identities they hoped to assume.

The scientists were also asked to talk about their doubts and what they perceived to be their limitations and weaknesses. I included questions about self-doubt because these would allow me to paint a multifaceted, and more complete and accurate, picture of how scientists see themselves. By including these questions, I proceeded on the premise that self-doubt often functions as a smoke screen for wish fulfillment. Doubts draw attention to those parts of a person's life that he or she wants to be better. Self-doubt may often result in compensatory behaviors that rectify or allay perceived deficiencies. Accordingly, this topic presents an additional lens through which to view aspiration, possibly a kind of aspiration that is driven more by feelings of insecurity than by resolute self-confidence.

I devoted one specific set of questions to ambition. In asking the scientists about their ambition (or lack of it), I intended to find out how it is important, if at all, and to whom. I emphasized ambition to uncover the forces that drive people in a profession. I thus hoped to describe the chronology of how ambition emerges and persists or fades over time and through different places. All of the questions I asked are listed in appendix A.

I requested that each scientist furnish a resume or curriculum vitae, and each complied with this request at the time of his or her interview. These are revealing documents. They are statements by the individual about the individual, or at least about those *parts* of the individual that he or she is required or wants to make known to selected others. They represent a kind of lifeline for most scientists. They bring, or fail to bring, varieties of rewards to the people who compile them. Consequently, they are normally crafted with exquisite care. They are also useful documents for the purposes here, establishing a history of where an individual has been and what he or she has done. They will be used to supplement the discussion of the narratives.

In addition, I asked each scientist to complete a questionnaire at the end of the interview. On the questionnaire, the scientists provided demographic information about their age, marital status, age of children if any, parents' occupations, racial or ethnic background, religion, and income. The scientists also provided information about their membership in professional organizations, their involvement in meetings and which ones, and their journal subscriptions. These latter questions were intended to provide a measure of occupational identification and commitment.

The questionnaire was supplemental. The narratives provided in the interviews will form the primary basis of my discussion. My use of information drawn from the questionnaire will be selective in the course of my analysis, since many of the questionnaire items proved to be only tangential to the major theoretic issues arising from the narratives. At various points I will note a scientist's age, marital status, income, or other pertinent information if it adds clarity to the account being presented. The questionnaire appears in appendix B.

The Universities

To investigate how self-identity may form and unfold differently in different types of places, I selected universities that on a number of counts stand far apart from one another. On first look, these institutions differ in prestige, but this was not the basis on which they were selected. In 1982, the National Research Council (NRC) assembled a panel to assess the quality of graduate programs in all doctoral-granting institutions across the United States. The findings were reported in *An Assessment of Research-Doctorate Programs in the United States* (Jones, Lindzey, and Coggeshall 1982).

This report evaluated departments according to six criteria: program size; the characteristics of graduates; university library size; the amount of research support; the scholarly reputations of faculty and programs; and records of publications attributable to the departments. Taking one of the indicators—scholarly reputability—I selected universities that scored at or near the top, middle, and bottom of all the universities that were assessed. I selected "scholarly reputability" as the key variable by which to differentiate departments because it represents the most general statement about departments and their members. In other words, it is the most *inclusive* of the factors that go into distinguishing one department from another.[2]

I selected top-, middle-, and bottom-ranked schools on the premise that differences in these institutions establish different conditions—both structural and cultural—for careers.[3] Characteristics of the schools, which I will discuss momentarily, attest to these different conditions. By including schools that represent the two extremes and the mean of environments in which scientists work, I have aimed to maximize variation in the careers of academic scientists.[4]

Universities as Social Worlds

Universities, and university departments of physics specifically, are distinguished from one another by a variety of factors. Some of these factors are the criteria that were used, as described above, to assess departments. The list of additional factors is infinite. Schools and departments differ, for example, by the size and scope of budgets and endowments, by the number of students and the degree of their selectivity, by the nature and number of teaching duties assigned to faculty members, and by the density of researchers in specialties.

Although each of these factors plays a part in shaping careers, either by opening up opportunities or by limiting them, a more general phenomenon sets institutions apart and encompasses a driving force in the ways that scientists form self-images. This is the *expectations* that environments hold for their members. Like law firms or city orchestras, universities differ by expectations of role performance. Different schools possess different norms of success, and the corresponding "ladders" in each school differentiate among levels of individual accomplishment.

The orientations that people bring to their work are both structurally and culturally constrained. Differences in the social structures of institutions make opportunities more available to some than to others (Crane 1965; Long 1978; Long and McGinnis 1981), a fact that is reflected in the scientists' appraisals of their own accomplishments. And there are subcultural distinctions in which people in different types of universities subscribe to specific beliefs about what constitutes achievement, extraordinary achievement, and lack of achievement. People who see themselves "at the top" in one environment may see themselves in the middle or near the bottom in others because the people in these places work according to different performance norms.

The cultural and structural differences among types of schools lend to their being portrayed as distinct social worlds. Entering each of the worlds is much like entering different parts of the world at large. In some respects the worlds resemble one another. In each of them one finds people involved in similar activities, using similar talents. But in other respects the worlds differ dramatically from one another. A scientist who subscribes to the expectations of one world would feel like—and be viewed as—an outsider in a world far removed.[5]

Worlds Together, Worlds Apart: Elites, Pluralists, and Communitarians

We need to give names to the distinct academic worlds in which scientists work, so as not to confuse one for another in the course of describing them. An easy but exceedingly oversimplified way to identify the worlds could be based merely on their status, resulting in a rough (and not terribly meaningful) vertical classification of high, middle, and low. It hardly needs pointing out that classifications emphasizing hierarchy abound.[6] But worlds differ on many dimensions, and the subjective views of those working in each setting do not depend entirely on any one or even all of those dimensions.[7]

Top-ranked schools correspond to the world best called *elite*. This is the world that places the highest premium on scientific research. "Elite" uniformly describes the members who work in this world and the external definition of them and their department. It also expresses the aspiration of its members—to be among "the best"—and the key collective goal that brings them together and establishes their membership in universities that are also elite. Some members may have second thoughts on, even strong dislike for, this aspiration as a personal or collective objective, but ultimately it is the sole standard that they unhesitatingly adopt.[8]

Middle-ranked schools correspond to the world best called *pluralist*. This world includes some members as eminent as those found at elite institutions, but the pursuit of still more eminence is not what holds members together, nor does it provide a standard that all members unhesitatingly adopt. This type of university answers to considerably more varied demands, those of mass teaching as well as research and service to the wider community and state. Service is apparent, especially in the application of the science and technology generated within its own quarters. This world needs to have eminent faculty members in its ranks in order to attract and maintain visibility and resources as well as to realize some success in what appear to be perpetual efforts at "institutional upgrading." But given the diversity of audiences to whom this world is accountable, the pluralist schools also need to achieve some balance among staff. Often this results in a blend of people who exhibit radically different affinities, talents, and motivations. As a division of labor, departments in this world mirror something of the "multiuniversity" of which they are a part.

"Tail" schools correspond to the world best called *communitarian*. The most defensible collective grounds for the actions people take are local, reflecting responsibility to the institution itself. Like pluralists, they answer to many demands—research, teaching, service, administration—

but the fundamental basis of comparative worth is within the institution itself. "Good citizenship" is demanded of all and is a primary basis on which individuals are accorded honor and esteem. All members are expected to shoulder more than one responsibility, and faculty almost never find ways to completely excuse themselves from teaching. Unlike elite and pluralist worlds, no special arrangements are made to attract "stars," those people who are known widely by academic as well as general public audiences. This world holds an ethic of shared responsibility.[9]

In this sample, elites are represented by two schools, pluralists by one, and communitarians by three. I interviewed a roughly equal number of scientists (about twenty) in each of the three types of schools.[10] The design of the study is summarized in table 1, which lists the number of scientists I interviewed at each type of school and within each of the three cohorts. Notice that the numbers of scientists are also roughly equally distributed by cohort.

Of the sixty scientists, four (or 6.6 percent) were women. In the United States, 5.4 percent of physics faculty members at graduate degree–granting institutions are women (American Institute of Physics 1994, see tables 1 and 2, pp. 2–3). Thus my sample represents a slight but negligible oversampling of women physicists.

The Empirical Basis of Social Worlds

"Conditions of practice" are essential to understanding the formation and unfolding of self-identity because they set the very stage for such processes. (In addition, conditions of practice further illuminate the bases on which each of the worlds assumes its collective identity.) Those that most directly affect people and their careers fall into the loose and varied category of "resources": people (such as teachers, students, staff); money; buildings; equipment; and the very nature of people's tasks. For nearly every resource, we may evaluate its quality and its quantity.

In table 2, I list a wide range of the most important conditions of practice that bear on science careers; these are differentiated by the worlds in which scientists work (several more distinctions will follow in the next chapter). As table 2 shows, resources—be they human, fiscal, or physical—are spread unevenly across the worlds.

Of the elite schools, one is a private university, the other a public state institution. Both have a long history of accomplishment in research, as well as in training large cadres of scientists to carry on in this tradition. All of their departments and programs have national reputations, and many command international audiences as well. Their departments of physics

Table 1
Number of Scientists, by Type of Academic Institution and Cohort

Institution	Cohort (by Year of Ph.D.)			Total
	Pre–1970	1970–1980	Post–1980	
Elite				
School 1	8	6	6	20
School 2	1	—	2	3
Pluralist				
School 1	6	5	7	18
Communitarian				
School 1	3	3	3	9
School 2	3	2	2	7
School 3	1	—	2	3
Total	22	16	22	60

are considered to be among the most distinguished in the world. A large percentage of their physics faculties are highly visible in the field, with a number of Nobel laureates within their ranks. The departments are among the largest of all U.S. university departments, as well as among the largest physics departments. Both departments have faculties numbering around seventy-five members. The private university often compares itself to and is normally identified with schools such as Princeton, Stanford, and Cal Tech, and the public university is akin to the Universities of Michigan and Wisconsin.

Pluralists are represented by a large public state university. This school has begun to enhance many of its programs and departments, some of which already receive national attention. The school is the state's flagship institution and thus devotes a major part of its mission to providing a college education for state residents. Although the school has long had a faculty committed to research, it has in the past fifteen years placed an even greater emphasis on this aspect of its endeavors. Ideally the school would hope to achieve the stature of public universities such as Michigan and California-Berkeley. Although members of the physics faculty see their department and university as having great potential, they freely admit that their department lands somewhere in the middle on a scale of reputability. The department is moderately sized compared with other physics departments, counting around fifty members. The physics department at this university is comparable to those at Indiana, Michigan State, and Iowa.

Communitarians are represented by the final three schools. These are smaller state institutions that started as "comprehensive universities."

Table 2
Conditions of Practice

	Elite		Pluralist	Communitarian		
	School 1	School 2	School 1	School 1	School 2	School 3
Faculty conditions[a]						
Department size[b,c]	≈75	≈75	≈50	≈25	≈15	≈15
% with Ph.D.'s from top-10 physics departments[c,d]	78%	74%	56%	16%	0%	1%
% with ext. funding[e,f]	≈50%	≈65%	≈50%	≈30%	—[k]	—[g]
% with grad. research assistant[h,i]	76-100%	—[j]	51-75%	26-50%	—[k]	1-25%
Reasons for no grad. research assistant[i,l]	no active research	—[j]	various reasons	no funding	—[k]	no funding
Number of postdocs[i]	16-20	—[j]	>25	6-10	—[k]	1-5
Number of technical support staff[i]	21-25	—[j]	16-20	—[m]	—[k]	1-5
Yearly teaching load (no. of courses)[j,n]	1-2	—[j]	1-2	2-3	—[k]	3-4
Teaching leaves?[i]	yes	—[j]	yes	no	—[k]	yes
Leave provisions[i]	negotiate with dept. head	—[j]	sabbaticals[o]	none	—[k]	sabbaticals[p]
Student conditions						
No. of undergraduates[q,r]	≈5,000	≈25,000	≈35,000	≈15,000	≈8,000	≈15,000
Avg. SAT scores[s,t]	680	570	515	480	485	510
No. of physics graduate students[c]	≈300	≈300	≈75	≈35	≈15[u]	≈5
Dept. GRE scores[c,t]	575 (min.)	740 (med.)	600 (min.)	600 (avg.)	580 (avg.)	550 (min.)
Avg. time to physics Ph.D. (years)[f]	≈7.0	≈7.0	≈7.5	≈8.5	—[g]	—[g]
% getting jobs at Ph.D. schools[v,w]	≈60%	≈60%	≈50%	≈45%	—[x]	—[x]
% getting jobs, general[w,y]	≈60%	≈75%	≈45%	≈60%	—[x]	—[x]

(continued)

Table 2 (*continued*)

	Elite		Pluralist	Communitarian		
	School 1	School 2	School 1	School 1	School 2	School 3
Other infrastructural conditions						
Annual fed. support to dept.[i]	≈$70M	—[j]	≈$7M	≈$800K	—[k]	≈$1M
Annual nonfed. support to dept.[i]	—[m]	—[j]	≈$1M	≈$450K	—[k]	0
Annual dept. operating budget[z,i]	≈$10M	—[j]	≈$7M	≈$80K	—[k]	≈$60K
Assistant professor start-up funds?[i]	yes	—[j]	yes	negotiable	—[k]	yes
Start-up amount[i,aa]	$40K–$400K	—[j]	$50K–$450K	≤$25K	—[k]	$60K
Restricted uses?[i]	no	—[j]	no	capital equip. only	—[k]	capital equip. only
Other dept. or university funds?[i]	yes	—[j]	yes	yes	—[k]	yes
Type/amount[i]	variable	—[j]	$10K grants; $4K aid(theory); $10K aid(exper.)	$5K grants; $1K–$100K (equip. grants)	—[k]	$2K grants
Travel/conference reimbursement?[i]	no[bb]	—[j]	yes	yes	—[k]	yes
Nature of provisions[i]	—[bb]	—[j]	≈1 conf./year	$250 dept. $750 univ.	—[k]	≈$500
Annual no. of outside speakers[i,cc]	>30	—[j]	>30	6–10	—[k]	26–30
Personnel policy (select issues)						
Outside letters for tenure review?[i]	yes	—[j]	yes	yes	—[k]	yes
Customary ten years ago?[i]	yes	—[j]	yes	yes	—[k]	yes
Customary twenty-five years ago?[i]	yes	—[j]	yes	no	—[k]	yes
Teaching as promotion factor[i,dd]	a lot	—[j]	somewhat	a lot	—[k]	a little

[a] Information applies to the *departments* of physics, except where noted. For certain matters, reporting of data is approximated (i.e., "≈") to ensure the anonymity of institutions and individuals.

[b] Number of *full-time* tenure or tenure-track department faculty members.

[c] Source: American Institute of Physics (1993).

d. Percentage of department faculty who received their Ph.D.'s from departments of physics ranked in the top ten by Goldberger, Maher, and Flattau (1995). Due to ties in ranking, the list includes eleven schools: Harvard, Princeton, M.I.T., Berkeley, Cal Tech, Cornell, Chicago, Illinois, Stanford, University of California–Santa Barbara, and Texas. To arrive at the percentage, I included in the denominator only department faculty members who had *U.S.* doctoral degrees.

e. Percentage of department faculty with outside research support.

f. Source: Goldberger, Maher, and Flattau (1995).

g. Not evaluated by Goldberger, Maher, and Flattau (1995) because departments did not meet criteria for assessment.

h. Percentage of department faculty with at least one graduate research assistant.

i. Source: Departmental Questionnaire (see appendix C).

j. Department head did not return questionnaire. Information on Elite School 2 most closely approximates equivalent information for Elite School 1.

k. Department head did not return questionnaire. Information on Communitarian School 2 most closely approximates equivalent information for the other two communitarian schools.

l. In response to the following question: "Taking all the faculty members who do *not* have graduate research assistants, what would you say is the single most important reason?" (See Departmental Questionnaire, appendix C.)

m. Missing data.

n. Does not discriminate between courses and course sections.

o. Also one-quarter reduced teaching load for special research assignments (according to department head, this is used by about 10 percent of the faculty).

p. Also unpaid leaves of absence (which must be approved by the department and the university); in addition, reduced teaching for administering large grants, which also must be approved.

q. Number of undergraduates in the university. (The number of undergraduate majors in physics was not available for all departments. More important, however, such figures obscure the extent of undergraduate teaching in "service" and other general courses.

r. Source: Barron's Educational Series (1996).

s. Source: American Council on Education (1992).

t. The maximum score for the Scholastic Aptitude Test (SAT) and for the Graduate Record Exam (GRE) is 800 for each of their components.

u. Includes only master's degree students; department no longer has a doctoral program.

v. Percentage of doctorates, 1975–79, who indicated they had made firm commitments for employment in Ph.D.-granting institutions.

w. Source: Jones, Lindzey, and Coggeshall (1982).

x. Not evaluated by Jones, Lindzey, and Coggeshall (1982) because departments did not meet criteria for assessment. Refer to note 3, this chapter.

y. Percentage of doctorates, 1975–79, who indicated they had made firm commitments for postgraduation employment (including postdoctoral positions and other appointments in academic and nonacademic sectors).

z. Excludes faculty salaries; refers exclusively to nongrant support.

aa. Amount depends on research program of specific faculty member.

bb. Presumably, faculty members rely on their research grants or unrestricted funds disbursed by the department or (probably rarely) personally cover their travel costs.

cc. Estimated annual number of people outside the department who gave formal talks.

dd. In response to the following question: "As best as possible, indicate the extent to which teaching performance factors into promotion decisions in your department." Answer choices: a lot; somewhat; a little; practically not at all; not at all. (See Departmental Questionnaire, appendix C.)

One of the schools was originally a teachers' college, and all three institutions began by granting only the first graduate degree—normally the master's degree—in programs that offered any graduate-level curriculum. The first two of these schools have had university-wide doctoral programs for about three decades and have achieved state and some regional prominence. The third established doctoral programs in the 1970s but lost the program in physics several years later as a result of state budgetary measures. The primary mission of these three schools is baccalaureate education, although they have attempted to change their identities by placing a greater emphasis on their research activities; the faculties of each of these universities are active in research and would be justifiably skeptical of any characterization that continued to depict their schools as essentially teaching universities. The influence of these schools in science is moderate compared with that of the other two groups of institutions but is by no means trivial. The departments of physics in the three schools are relatively small. Two of the departments have about fifteen faculty members, and the third has about twenty-five. They are akin to schools such as the University of Tulsa, Drake University, or a branch campus of a university system, such as the University of Wisconsin–Milwaukee.

Nota Bene

Different worlds of science, like different parts of the world at large, need not be viewed as "better" than others, despite the popular and sometimes invidious distinctions that are made among universities that differ in prestige, power, wealth, or public visibility. Worlds of science need only be viewed as *different* in how people orient themselves toward specific sets of values and beliefs and in how people adapt to available opportunities. An industrialized country can appear misdirected and uninhabitable by someone whose life is organized by a different economy, just as a university that is assigned considerable prestige can appear unattractive and inhospitable to someone whose priorities or ways of work differ. One person's misery is another's paradise, and I make this distinction to alert readers of the need to understand that people's careers are described here in terms meaningful *for them.*

Although the worlds bear certain similarities—all are, after all, universities—it is their differences that must be highlighted for comparative purposes. It thus may be easy for a reader to forget how sociology is being used here. Because my discussions will be based on comparisons, and because my topic deals with groups to which academic readers are themselves intimately tied, there is likely a built-in temptation to interpret the discourse as a form of criticism.

Resist that temptation. I am using sociology to understand how identities—both individual and collective—are framed and built. This process necessarily involves contrasts between different types of places and between people who do different things and possess different values, norms, outlooks, and ambitions. To speak about the characteristics of one world, then, is not to speak about them in absolute terms but in terms relative to what is found in the other worlds.

Methodological Considerations

The use of any research method or any combination of methods raises a number of issues pertaining to the strengths and limitations of the methods in yielding qualified truths about the phenomena studied. The use of personal narrative is, of course, no exception. The narrative approach to the study of social life was in large measure instigated by W. I. Thomas and Florian Znaniecki's classic five-volume study *The Polish Peasant in Europe and America* (Thomas and Znaniecki [1918–20] 1958). Thomas and Znaniecki's study of the assimilation of immigrant groups pioneered the use of personal "data" sources as a major method of sociological inquiry.[11] Life histories, diaries, letters, notes, certificates, newspaper columns, and editorials became primary bases for exploring the relationship between individual and society. In particular, these sources cast light on how people make sense out of and create order in their environments and on how they respond to social change. Thomas and Znaniecki's study and those that followed its lead spawned a tradition of work known as *biographical approaches* to the study of society, of which personal narratives, life histories, life stories, case histories, and autobiographies are representative parts.[12]

The strengths that characterize biographical approaches are those that distinguish qualitative research more generally. Their hallmark is depth of detail and analysis. If one sets out, as I am here, to investigate how people make meaning of themselves in relation to the larger social order of which they are a part, the method of inquiry calls for an ability to differentiate among fine categories of meaning in which people construct and locate themselves and others. The processes by which this meaning is created are ultimately discursive (Cooley 1902, 1909; Mead 1932, 1934; Wiley 1995). The analysis of how self-identities are formed, therefore, must focus on narrative, for it is in the stories that people tell themselves and others that we understand how individuals repeatedly arrive at some measure of self-understanding (see Brooks 1984; Mitchell 1980).

Nevertheless, it is important to highlight the most salient constraints to a narrative approach. This exercise in self-criticism is by no means intended to dismiss methodological issues surrounding narrative analysis. Rather, I am attempting to provide the reader with an informed stance toward the data and my interpretation of it.

Coherence

In an influential essay, Bertram Cohler (1982) observes the social basis on which narratives acquire consistency and stability. Cohler argues that stability and consistency over time is due more to reconstructive maintenance of personal narrative than to constancy of development. "The form of narrative is based on a socially shared expectation that stories have a beginning, middle, and end" (Cohler 1982, 205). The failure to maintain a coherent personal narrative, according to Cohler, results in feelings of fragmentation and disintegration.

Cohler argues that the life course is marked by several transformational periods in which personal narrative is subject to change in response to a combination of "intrinsic developmental processes," socially influenced conceptions of an individual's place in the life course, salient life events, and historical influences on individual lives (Cohler 1982, 214). One such transformational episode occurs generally, according to Cohler, at entry into middle age. Seen as part of the larger life span, this phase is especially key to the study of adult working professionals. The feature of greatest analytical interest as it bears on how narratives are told is reminiscence, or the process of recall and remembering (Rubin 1986; Unruh 1989).

"The reality of a limited lifetime is increasingly apparent across the middle years. The death of parents and close friends serves as a constant reminder of mortality. However, mourning is necessary not only for such losses but also for lost hopes and aspirations" (Cohler 1982, 223), for what has been described as the "dream" (Levinson 1978). "*Grief work* makes possible increased reconciliation with life as lived, including disappointments with the past" (Cohler 1982, 223, italics mine). "As a result of increased awareness of the finitude of life, reflection upon the past . . . becomes a central theme in the organization of the personal narrative across the second half of life" (Cohler 1982, 224). The middle-aged and the elderly, loosely defined, use reminiscence differently: the elderly attempt to settle accounts before death, whereas the middle-aged attempt to solve problems in the present (Cohler 1982; Marshall 1980).

Consistency in personal accounts is rendered problematic by the change that people experience as their lives unfold, whether this change be individual, social, cultural, and/or historical. Like Cohler, Charlotte Linde (1993) seizes on the issue of coherence in narrative accounts. In reflecting on their lives, according to Linde, people face a disparate, sometimes troublesome set of events and fateful turning points, many of which only in hindsight may appear consequential. When attempting to interpret these events and turning points, an individual is expected to create "a coherent, acceptable, and constantly revised life story" (Linde 1993, 3). For Linde, the life story consists of all the accounts about an individual's life that have been told by that person (to himself or herself or to others) over a lifetime (or some portion thereof) and that currently belong to the individual's repertoire of stories. A life story is thus an interpretative assemblage of countless events, experiences, and feelings, all of which are viewed as informative parts of a whole. But note the provision for selectivity. A life story is a *reconstruction,* subject both to deletion and to revision, whether deliberate or unintended.

Thus, whereas the personal narrative may be seen as an attempt by the individual to achieve and maintain coherence in light of life's complexities, the narrative stands as a stylized presentation of self. Orville Brim and Carol Ryff (1980) identified four fallacies often made by researchers who use life events to explain behavioral processes. One is *vividness:* people often seize on a dramatic event as a cause for some change or effect. A second is *recency:* people often select the most recent event as cause for some outcome. A third is *size:* people think big events have the most power, and thus they tend to overlook the gradual, cumulative impact of small events. A fourth is *simplicity:* people tend to attribute cause to a single event, ignoring the interactive effects of a number of events (Brim and Ryff 1980, 384–386).

Metainterpretation

Several fallacies relate to causality in and validity of narrative accounts. How is one to believe that the ordering of events as reported is an accurate portrayal of their sequence and timing? And how does one have faith in an account that by definition is a selective, perhaps patently false, representation of a life or some aspect thereof? A respondent's account is necessarily biased by time, memory loss, and the deliberate deletion or insertion of material that stylizes the account.

This study engages in metainterpretation. It interprets how people explain their passages in a career, specifically how they generate meaning about themselves within the social systems in which they and others are enmeshed. My objective, then, is not to assess how accurately individuals recall the past, for that would be nearly impossible. It is rather to see how individuals tell stories about themselves, where they assign meaning, how they organize their complex pasts to present a coherent self-identity.

We should, then, view the accounts that I present here as constructed representations of individuals by the individuals themselves. The accounts are, in principle, causally correct and valid. That is, the objective aspects of the accounts can be checked. Did Mr. Smith really attend the University of Connecticut as a graduate student? Did Ms. Johnson really become a professor in 1974? Did she hold a postdoctoral appointment at the lab where she claimed she did, and did the appointment truly last for three years? Such features of an account can be verified, and in fact, having each individual's curriculum vitae renders much of this exercise uncomplicated.

The subjective aspects of an account—the narrator's remembrances, feelings, emotions, and so on—are a somewhat different matter. They are stylized aspects of self-presentation and for this reason should be critically examined. Is this really how Mr. Smith felt when he lost his job? Is Ms. Johnson being honest when she says she is happy with the way things have turned out? Many such questions were scrutinized during the interviews so that the interviewees provided, as well as they could, a rationale for what they were saying. That is, what people said had to make sense. Their accounts are, or strive to be, internally valid.

The internal validity of accounts informs their external validity—the broader degree to which the accounts are true. I make this claim on the following premise: in the context of an in-depth interview involving a series of probes and conducted over an extended time period, it would be difficult for a respondent to sustain falsehoods or parts of a narrative that are in some way unpersuasive or that do not make sense. Given the length of the interview and the persistent probing that occurs, respondents are held accountable for what they say. Other than by declining to answer a question (which never once occurred in all sixty of the interviews), the scientists were to respond in a manner that satisfied the way in which the interview situation has been defined, namely as an occasion to talk candidly about themselves.

Even if we knew someone was lying or in some other way providing an account that did not ring true, that too would tell us something about the person and about the social contexts in which that individual is and has been a part. Moreover, if we knew that several people were lying and if we could identify patterns in the ways they lied, that would also establish a ground on which to generalize about people, their environments, and the situations they encounter. Viewed in this way, validity becomes a moot issue. The objective here is to take what people choose to say about themselves, and how they say it, and to abstract, successively, generalizations about social contexts of scientific work and the people who belong to them.

The Few and Far Between

Finally, the scientists I am studying are all relatively successful, in the sense that they have continued practicing their craft. This sample does not include what might be called the "losers," those who dropped out after their first brush with academia. In an examination of several cohorts of doctoral students at a subset of elite physics departments (where a majority of the physicists in this sample earned their Ph.D.'s), William Bowen and Neil Rudenstine (1992, 400) found that fewer than three out of four students who entered a physics Ph.D. program went on to receive the Ph.D. degree.

For many of those who do earn a Ph.D. in physics, however, the story is even more grim. In 1973, the percentage of physics Ph.D.'s who ended up working completely outside of physics stood at 32 percent (Kirby and Czujko 1993). This percentage rose over the subsequent decade of retrenchment in U.S. higher education such that by 1989, the last year for which data are available, some 43 percent—close to half of all new physics Ph.D.'s—entered some kind of nonphysics employment (Kirby and Czujko 1993). Taking a job outside of physics is not usually the primary motivation behind getting a Ph.D. in the field. At the same time, viewing such people as "losers" is premature and is the view taken only by those on the upper floors of the ivory tower when someone leaves work like their own and is never heard from again. For some people, an exit from physics or from science altogether may be a blessing in disguise.

I make these points because they bear on ambition. It is doubtful that those whose careers are fundamentally blocked lack only ambition. Talent, ability, creativity, perseverance, serendipity, and other factors come into play. Conversely, some scientists may carry a surplus of ambition but be denied access to academic physics simply for want of jobs. It is not uncommon for a department of physics—any department of physics—to receive three hundred or more applications for one vacancy.[13] My point is that most (if not all) scientists in this sample possess (or have in the past possessed) ambition of some kind. If they did not, they most likely would not be in the positions where we find them today. This is not a study about who has ambition and who doesn't. It is about how ambition of various types and intensities plays out in a career—emerging, flourishing, and dying by virtue of the structural and cultural conditions that mark worlds of work. And it is about how individuals react to these continuities and changes, crafting self-identities that respond to a complex past in light of a desired future.

TWO

Academic Worlds

People's beliefs and actions do not arise or manifest themselves in isolation from the places in which people interact. Universities provide scientists tools with which to craft a self-identity, much as environments do for other occupations. Situations and their consequences are defined as real through collectively generated systems of meaning (Garfinkel 1967; Goffman 1959, 1967; McHugh 1968; Thomas 1923). To eventually place the accounts of individual scientists in their proper contexts, therefore, we need to acquire a detailed understanding about the group modus operandi found in each of the three types of universities where scientists work. In this chapter I will outline the moral orders that establish guides for action and a ground on which individuals build their self-identities. In doing so, I will draw a profile of what scientists take to be normal and expected in the environments in which they work. I will pay special attention to scientists' "ways of knowing": that is, how they envision and go about behaving in their academic worlds.[1]

Erving Goffman spoke of the "moral career" in his explanations of how people manage social stigmas or "spoiled identities" (Goffman 1961, 1963). The moral career refers to the identifiable sequences in a labeling process in which a person's identity and moral status are progressively changed. The mentally ill or the physically deformed undergo a process whereby their identities form and evolve in light of what a wider group or society defines as "normal." Goffman noted, "One phase of this socialization process is that through which the stigmatized person learns and incorporates the standpoint of the normal, acquiring thereby the identity beliefs of the wider society and a general idea of what it would be like to possess a particular stigma" (Goffman 1963, 32).

The concept of the moral career has broader applications. It may be applied in this instance to groups of people who, in some degree, are personally invested in their work. As it pertains to scientists, the moral career refers to how a group defines a normal set of status passages (Glaser and Strauss 1971). The passages may be few or many, depending on the work environment, but they always involve turning points around which individuals modify their self-identities in light of newly occupied statuses. The moral career embodies a generalized perspective. Subjectively, it prescribes sets of beliefs, dispositions, and practices that people adopt in order to satisfy specific status expectations. Individuals render personal judgments about themselves by viewing their own careers against the backdrop of what is accepted and viewed as normal by the wider group. As scientists age over time, their entire personal biographies may be reinterpreted in light of how they see themselves standing with respect to collective sentiments about how the career should look at progressive stages.

In each of my interviews, I asked the scientists how they conceive of success and failure. I asked what "ultimate" success means and what they associate with "minimal" success. I further asked the scientists to describe, as best they could, how the successes of their local colleagues vary. These questions were not meant as exercises for scientists to speak about careers in oblique abstraction. Rather, they served as ways for them to characterize careers, indeed varieties of careers, by placing the careers in a cultural context. In talking about how scientists conceive of success, of extraordinary success, and of lack of success, we are talking about two central aspects of career experience. One is how personal *legitimacy* is established (and sometimes lost) within specific social environments. The second encompasses *images* of how the self is modeled. In each of the worlds, careers are symbolic: they stand for something greater than the individual. In seeing how scientists depict their own and others' careers, we canvass a cultural landscape in which those careers acquire meaning and significance.

Mapping Moral Orders

A moral order, as the term implies, consists of a social arrangement of desired and valued actions, beliefs, and orientations held among a group that has been organized around a certain purpose or set of purposes. In other words, it specifies how things *ought* to be (even though reality almost always deviates from the moral prescription).

Of the three academic worlds, the elite is distinguished by the extreme care that scientists gave to discussing careers that pass through it. They normally described, with exacting detail, the moral career that all in the elite world routinely strive to follow. Careers of elites, more so than those of pluralists or communitarians, are institutionalized: there are many socially established and expected phases.

The elite careers are standardized and explicitly hierarchical. Symbolically, they are seen to mature and prosper as they unfold, growing ever more advanced and supreme. People also are cast in successive roles in which they embody mounting worldliness on a path toward eminence. To be sure, elites do not always adhere to the moral career. We see in this world internal variety as well. But they possess an unrivaled conformity in beliefs about what constitutes desired performance and how that performance should evolve over successive phases of life.

Elites are relatively homogeneous in their commitment to and identification with science. All are, and have been throughout their careers, active in research. Most centrally, their commitment and identification is socially controlled by those with whom they work. Those who falter in the course of their careers expose themselves to feelings, if not socially enacted rituals, of degradation. When individuals fail to comply with minimum group expectations by halting their research altogether, severe social sanctions (not to single out economic sanctions) are normally imposed, including isolation and loss of power. In the elite world, scientists normally work with the assumption that individuals are only as good as their last work.

The reality in which scientists find themselves is different in the other academic worlds (cf. Berger and Luckmann 1966). The spectrum of commitment and identification broadens as we move into the worlds of pluralists and communitarians. Where the spectrum is broadest, among communitarians, the commitment to and identification with science is most varied and uneven. Scientists in these institutions are heterogeneous in their beliefs and practices about what defines a legitimate career. Those who lead essentially teaching careers, or careers in which research has been sporadic over the course of time, are most likely found here. In contrast to the conformity of the elite careers, diversity characterizes the careers that communitarians socially certify as valid. They are accepting, or at least tolerant, of a wide latitude of scientific commitment. In accounting for the way in which individuals establish legitimacy here, this is a world in which scientists believe that the person comes before the work: individuals are respected on the basis of their human virtues.

The remaining world of science, pluralists, represented by the large public state institution, falls in between the other two types. Here one finds scientists whose identification and commitment resemble those of elites alongside scientists whose beliefs and practices are more closely allied with those of communitarians. There is a tolerance for a diversity of practices and engagement with science, but the span of this tolerance is not as wide as among communitarians or as narrow as among elites.

Elites

The moral orders of the worlds are more fully elaborated by those who make their careers in these worlds. Scientists who describe what it is like to work in these worlds shed more light on the baseline conventions from which people operate. Our first informant is a young scientist who recently earned tenure at one of the elite schools. He works at a public institution with a highly reputable department of physics. For this scientist, the names of other scientists at the school symbolize many of the qualities that the institution stands for.

> When I walk down the corridors of this institution and I look at the names on the doors, to me, it's like being in the hall of fame. The names that one sees on the doors, for me, are the names of heroes.[2]

The influence of place on the person is keenly felt. Places transmit values that frame one's way of behaving and knowing. Our informant talks about his career in its environmental context:

> It [the place] has really benefited it [the career] because the standards are just uniformly high, and I'm not convinced that young scientists have terribly well-defined intrinsic standards. I think the very best may, but generally speaking, I think people's research quality and attention to detail is to a substantial extent colored by the expectations of their colleagues. I'm not saying that people scrutinize my papers. I can send out whatever I want, but my colleagues are just uniformly high-caliber people, people with really high standards of research and integrity and commitment to the [science] community. And because of that, I know that's how my parameters got set. I would hope to be, for the way I conduct my research, the part that I have control over, I would hope that those kind of high-caliber characteristics sort of rubbed off. You swim with Olympic swimmers and you behave like one.[3]

Environments also constrain. In a world of Olympic swimmers, in which all or most of the competitors seek the gold, the most daunting constraint is the place itself. The call to success operates not only as an incentive but also as a burden.

There is one constraint, and that is—this sounds absurd-but in a way, *success* is a constraint. Everybody here pretty much is successful, to the extent that they are regarded as international leaders in whatever field they choose to work in. What that means is, it means that sometimes you have to, you feel you ought to continue to produce in something, in an area in which your interest is slightly dwindling, in order to maintain a high profile. No department head has ever said, "You should continue to work in such and such an area." Nobody has, except colleagues in informal discussions with no sort of scrutiny, but just sort of friendly chats. We all talk about where we are headed and what we are doing, how we should best foster our own careers. . . . To some extent, the quality of the place is a responsibility.[4]

At root, the academic elite is a world of activity. People appear busy. They are involved in matters they view as important. There is an air of seriousness and dedication. Every world displays gradations of activity, but as we see, even what is considered the barest of effort in the elite world still involves engagement in the same activity—research—though in reduced magnitude. The moral order includes a collective understanding of what one should do minimally to maintain respect.

You have to publish papers on something in the mainstream journals. And it's absolutely clear what they are. If you show me somebody's publication list for the past five years, just by looking at the journals, I can tell you whether they are active or not. . . . *Physical Review, Physical Review Letters, Journal de Physique, Journal of Physics,* and many more. There are learned peer-reviewed journals, and then there are the rest. And if you are not publishing at least three papers a year . . . in those sorts of journals, then you have ground to a halt.[5]

Moral orders may be viewed as two sides of a coin. We learn as much about groups by seeing what they desire and expect as by ascertaining what they shun. Worlds of science, like worlds of other occupations, establish boundaries of performance. Members of worlds possess conceptions of success as well as of failure. These conceptions are defined on both social and individual levels. Personal definitions of success and failure may differ among individuals, even among the individuals who compose one group. Scientists may be viewed as "successful" by dint of having been rewarded by the science community. At the same time, those same scientists may view themselves as failures for not having satisfied personal expectations set at a higher level. In spite of such individual differences, people recognize the social definitions that distinguish among performance.

For elites, we have heard where the minimum is drawn: writing each year a few papers that are published in select journals. For elites, failure is drawn close by (but this line does not extend into other worlds, where members define performance differently).

There are people who just stop publishing and don't do anything to fill the gap because, even though we are university professors, the notion of teaching, which is what we are paid to do, is much more than standing up in front of a classroom. That only is what we do three hours a week. If you do it well, it probably costs you six or seven hours of preparation for each of those one hours, at least the first time you teach a class. But that still leaves twenty or thirty hours a week for research, and so teaching means being interactive with your graduate students and your post-docs. If you stop doing that, and you do nothing but your classroom teaching, then no matter how well you do that classroom teaching, you are a failure. If you assume some other responsibility, then you are acceptable. If you don't, then you are essentially ostracized. Most people here continue to be successful right up until retirement, continue to be productive, thoroughly productive. Being sixty is not really a disability. There are plenty of wonderfully active sixty-year-olds. If you take on a responsibility like getting involved in university administration, that's fine. If you do nothing, you are considered a disappointment, a failure.[6]

Elites describe the moral evolution of careers in ways that highlight gradations of achievement. The expectation is one of continuous productivity, even though there may be intermittent and alternating periods of the fallow and fecund. Within the moral order of elites, the person and the career look different at different times.

From graduate school to death, there's a sort of continuum of accomplishment, and what happens to people is that they rise as they age; they rise at different rates and then they stop somewhere. That's probably true for everybody, but they don't come down again. You pretty much have what you have done and then you are settled at some level; you can rise at different rates, and then you check out at different levels. We have here people who are about as successful as you can be. I'd say of the sixty to seventy faculty here, there are about ten . . . who are in their late fifties, early sixties, who are as successful as they can be, except they haven't got the Nobel Prize. So they are the kinds who are members of the National Academy of Sciences, and every year they get their membership to some new club of sorts. They are regarded as the elder statesmen of physics. They are all men; they all sit on national panels, advisory panels, all that sort of stuff; they spend time in Washington. By and large, they do very good research as well. Then there are people who should rise into those positions. Those are the people who are in their forties. Most of them are still at the stage in their lives where they haven't, they can't rest on their laurels; they still have to keep going. The sixty-year-olds are tremendously friendly, and they have us over to dinner, and they make a fuss of our kids, and they are supportive, and they just assume that we are like them—that we, in our early thirties, are just clearly the next generation of them. I do that with undergraduate and graduate students. I just assume that they are also going to have successful careers in physics. I wonder if one is not discerning, I don't know. The ones who are younger are a funny breed. They are quite tough, and they are still a bit competitive, and they are not so easy to live with. They are a bit more argumentative. I guess they are still trying to make their reputations, and so they are not as avuncular as the older people.[7]

The career is thus symbolically divided into age-graded statuses. An image exists of how the career unfolds. This image involves the objective statuses that are part of an evolving sequence of growth and development, in addition to the dispositions and manners that accompany each stage.

Let's start from the end. There's the landing, where you keep your copybook clean. These people that we work with are quite tough. There are the people who are the superheroes, the National Academy and that sort of stuff. They never really retire; even if they officially retire, they are still active physicists in the community, and people perceive them as such; and they get their invitations to talk at international conferences, and then they get to the stage where they are on the advisory committees, so they don't even speak, but they decide who does speak. And then earlier than that, that same bracket of people will typically be tremendously industrious in their late thirties to mid-forties. And so there's the pre–national committee advisory stage where they are making their reputations by doing really beautiful work, and that often involves very deep reflection and frantic work and complete submission to the problem. Everybody accomplishes something at some level and everybody finds his own community or her own community. The question is how rarified is that community.[8]

Communitarians

Communitarians operate on terms far different from those of elites. These scientists do not conform to any single model of a career, especially in their beliefs but also in their practices. This world of science is marked by a tolerance for varieties of identification with and commitment to science, whether primarily as teachers, researchers, or some combination of the two.

> I am very, very happy. I think that I have been able to function primarily in a teaching role. I am not a big researcher. I am able to do some of the type of stuff that I like dabbling around with, and it's not stuff that's going to be in *Physics Review Letters* [sic] or anything like that because it's not breaking news. To me it's like puzzle solving. I've been able to teach and do my little things.[9]

Some of these scientists view the terms under which they and their colleagues work in historical context. They highlight some of the changes in their world regarding how a group defines expectations for its members.

> The typical is changing. It's getting tougher in terms of the criteria. The criteria are increasing. Whereas in the seventies when I came here, the late seventies, the criteria were to get some token outside support to train a few graduate students, to do a good job teaching. It's always been to do a good job teaching. Do some amount of service, be on some committees and do some service for the department or the university. And then you have a successful career. I don't think you can have, I don't think you can base your career success just upon the evaluations that you get here. I think it's bigger than that. For me to be successful I think I not only have to [wow] my colleagues here, but I also have to be thought well of by the outside community in physics, I think. That would be more satisfying to me.[10]

Changes in expectations brought about by market forces and by the evolution of school missions produce cynicism in those whose beliefs about their roles differ from the organization's. This is most frequently found in people who age in ways incongruent with the way in which their place of employment "ages." In drawing lines between types of performance, one communitarian depicted how "ultimate" achievement is construed. But notice how both the scientist's tone and the actual threshold differ from the tone and threshold of elites.

> A lot of people would differ with me on this, but to some people, including some of those in this department, a successful person is one

who is aggressive and goes out and gets grants and establishes big groups, writes lots of papers, and is involved in all kinds of activities. That's a successful person. Well, I wouldn't go quite as far as that, because a lot of stuff is just aggrandizement: big grants, makes a big show, looks good, looks impressive and a lot of things running around, a lot of students running around, but I would not necessarily say that was a successful program, unless the work was of some significance. If it's really contributing to an understanding of nature, not just a window-dressing operation, not just, here, we've got a piece of equipment and we can run all this stuff through this equipment and get all kinds of results out and just write paper after paper like a machine or a mill, a factory, a paper factory.[11]

In the following depiction of "minimal" success, notice again the differences in tone and terms that distinguish the communitarian world from its elite counterpart.

Someone has to be at least diligently working in whatever endeavor they are pursuing, it might be teaching or whatever and really trying to improve and all that kind of stuff. Or if they are doing research work, to be not goofing off and playing golf and fooling around like that. Really putting in honest effort, I would say would be the minimum expectation.[12]

The acceptance of wide disparities in performance is associated with an occupational subculture in which people, apart from their work, form the basis for respect.

There are some people who do a wonderful job teaching. You cannot do everything perfectly. And so we have some people spending most of their time as excellent teachers. We first respect [people] for who they are. They are all nice people. If they do a very good job at teaching, they are respected, and if they do a very good job at research, they are respected. So they get respected for whatever they do, and some do both.[13]

Pluralists

Many of the distinctions that characterize communitarians also characterize pluralists.

Someone I admire in the department is named———. He's someone who did a significant piece of work back in the fifties and really hasn't done anything since. That's why I pick him as an example. If he had aspirations to be a leading theoretician, then they weren't fulfilled. Nevertheless he's an admirable teacher and colleague. He's very happy figuring things out and explaining things. So I would see his career as very successful. He's very admired in the department.[14]

But in the pluralist world, as the name itself indicates, one also is apt to find people whose sense of a science career resembles that of elites, to the extent that their narratives underscore research. The exact ways in which research in the pluralist world interleaves with teaching may not correspond directly to the research/teaching relationship among elites, since pluralists (those at large state universities) more often have sizable and powerful audiences to whom they are accountable as teachers, whether those audiences are students, university administrators, or legislatures. Nevertheless, several pluralists highlight the importance of research careers in ways akin to elites.

I think any of the things I've done—all would be successful. If I remained working in environmental technology, I think that's a successful career for a physicist. If I did applied research at an industrial laboratory, that's a successful career. If I was an administrator for the university, that would be a successful career. In carrying out a research program in physics, which is what people more traditionally see as a successful career, that's also successful.[15]

A group risks becoming polarized and dysfunctional when its members have sharply different affinities. Communitarians and pluralists are able to function collectively despite inside differences in commitment because of their style of interaction. In having the person be the prime basis on which to accord respect, these groups move the person's work away from center stage (except of course in private formal evaluation). In doing so, they create a social ground on which people who have different affinities can operate in relative harmony.

Places, People, and Performance

Our informants have identified several activities that compose the careers of those who are making their way in their respective social worlds. Teaching, advising, and training students constitute a key part of these activities. Among elites, however, these activities are treated as a constant within a larger rubric that encompasses steady growth in the scientist's identity as a researcher. Teaching and advisory roles are often performed in conjunction with the scientist's specific area of expertise and current research endeavors. This turns our attention to the research side of science careers.

Table 3
Journal Publications among Scientists, by Academic World and Cohort

Academic World	Cohort I (Ph.D. Pre–1970)	Cohort II (Ph.D. 1970–1980)	Cohort III (Ph.D. Post–1980)
Elite			
N:	8[a]	6	8
Maximum:	272	183	66
Minimum:	44	30	12
Mean:	109.3	91.3	29.3
Pluralist			
N:	6	4[b]	7
Maximum:	89	152	37
Minimum:	19	33	13
Mean:	60.7	73.3	23.9
Communitarian			
N:	7	5	7
Maximum:	100	62	32
Minimum:	9	6	10
Mean:	39.1	30.4	21.7

Source: Curricula vitae of scientists who participated in the study. The number of publications for each scientist includes all published journal articles. The number excludes the following: books; textbooks; book chapters; edited volumes, conference proceedings; invited and contributed papers; book reviews; encyclopedia, world book, and yearbook entries; and articles listed on the individual's curriculum vitae as "submitted," "in press," "accepted for publication," "in preparation," and so on. If the same journal article was published multiple times, it is counted only once.

[a]Data was not available for one case.

[b]One case was excluded. This case is treated as an outlier because most of the scientist's career was spent as an industrial scientist and thus does not represent the normal conditions under which the publication records of academic scientists are established.

Several tables of numeric information illustrate the research conventions prevailing in each of the worlds. First, table 3 gives the number of journal publications among scientists, listed by academic world and by cohort. Notice that in each world, scientists begin (Cohort III) with roughly equivalent records of journal publication. The most notable difference is between elites and communitarians, where the mean differs by an average of 7.6 papers. The greatest difference involves the upper limit, where elites outpace their most advanced communitarian counterparts by a substantial margin (66 papers as opposed to 32).

The more noticeable differences across the academic worlds occur in careers over time. The research careers of elites advance. Steady increases in the records of publication mark the careers of elite scientists in the middle and eldest cohorts. In the pluralist world, the grip of commitment is not as tight among members of the eldest cohort. Before this time in the career, records of publication have peaked. Comparing the eldest cohorts of elites and pluralists, notice not only the difference in the mean number of publications but also the range. The maximum among elites (272) is vastly greater than the maximum among pluralists (89). The same pattern is found at the minimum boundary (44 papers as opposed to 19).

The patterns found in the third set of institutions—the world of communitarians—underscore its differences from elites and pluralists. Although there is an increase in activity across cohorts, the change is comparatively minimal. On average, the eldest scientists have produced only 17.4 more papers than the youngest. And the minimum number of publications in the two oldest cohorts is actually *less than* the minimum in the youngest cohort. This pattern is likely part of a historical trend. It suggests that less prestigious institutions are benefiting from an influx of scientists who demonstrate a greater commitment to research than that of their older colleagues. This most likely stems from a combination of factors, including an increased emphasis on research in all types of institutions, even those that have traditionally had teaching-oriented missions, as well as a contracted academic labor market that has channeled research-oriented graduates to the periphery of doctoral-granting institutions.

The information presented in table 3 is consistent with the findings of Paul Allison and John Stewart (1974). Based on a cross-sectional survey of chemists, physicists, and mathematicians, their research shows that highly productive scientists maintain or increase their productivity, whereas less productive scientists produce even less as their careers unfold. They explain these patterns in terms of "accumulative advantage": feedback through recognition and resources successively benefits the productive while lack of such feedback successively impoverishes the less productive. The distribution of productivity becomes increasingly unequal as cohorts of scientists age. In plainer words, "the rich get richer at a rate that makes the poor become relatively poorer" (Merton 1973a, 457).

One could raise the counterargument that these distinctions in academic worlds are caused in part by people making, or failing to make, their marks elsewhere. In effect, this argument claims that the positions of people stem from past performances in different worlds. Thus, elites (or at least a fraction thereof) have gotten to where they are because of strides they made as former pluralists; pluralists inhabit their world because of their accomplishments as former communitarians or because of their failings as elites; some communitarians may have entered their world after failing as pluralists, and so on.

We can quickly dispense with this reasoning, however. Table 4 confirms that *nearly all* of these scientists entered and stayed in the same worlds in which we now find them. Only four elites held positions in

Table 4
Scientists' Mobility Patterns between Academic Worlds

Present Academic World	Formerly Elite	Formerly Pluralist	Formerly Communitarian	Formerly in Industry	Total Mobility Cases
Elite[a]	3	2	0	1	4
Pluralist[b]	0	1	1	1	2
Communitarian	0	2	3	0	5

Note: In the first two rows, numbers of "former" positions do not equal the total number of mobility cases because one or more of the scientists held more than one position.

[a]In one case, a scientist held positions first in the pluralist world and then in industry before entering the elite; in another case, a scientist held positions first in the elite and then in the pluralist worlds before returning to the elite.

[b]In one case, a scientist held positions first in the communitarian world and then in a pluralist world before assuming his current position in a different pluralist world.

other universities (or in industry). In three cases, the other universities (Princeton, Carnegie-Mellon, and Cornell) and the one industry (Bell Laboratories) were elite. Only one pluralist held a position at different universities (Duke, a pluralist institution verging on elite, and William and Mary, a communitarian institution). Only one pluralist previously worked in industry (at Xerox). Only five communitarians held positions in other universities, and in three cases the universities were communitarian. Thus, when it occurred, most mobility was between similar types of environments.

Of the eleven scientists who moved, eight brought tenure with them to their current positions. Thus, of the sixty scientists in the entire sample, a vast majority (52) have earned, or are attempting to earn, tenure in the settings where we now see them. The ways in which scientists' careers have developed or failed to develop are therefore rooted directly in their immediate organizational contexts. Thus our further analysis need not be encumbered by notions that achievement or lack of achievement has been followed by mobility across these academic worlds (except for the few cases documented). Mobility between academic worlds—even between institutions within one type of world—is low.[16]

The number of publications as an indication of the work done in academic worlds should not, of course, be taken at face value. To be sure, quantity is not the same as quality. As for quality, the moral orders of worlds are differentiated further by taking into account where physicists publish their work. Journal articles—as opposed to monographs, books, and other outlets for scholarly writing—represent the most common venue for published scientific work in physics and for most of the physical and biological sciences more broadly.

The number and range of journals to which physicists submit their work is best described as astounding. In 1973, across all subfields of physics, there were 277 journals in which experimental, theoretical, and pedagogical work could be published. By 1993, the latest year in which the most complete count could be made, this number had grown to 410, an increase of 48 percent, or roughly seven new journals in each year of that twenty-year span.[17]

Members of the physics community recognize, however, that only a tiny fraction—approximately 5 percent—of that total consists of journals that are regarded as "major" or "mainstream." These are known as the "leading" journals, all of which are refereed by colleagues in the community and which report what the community (or at least a powerful subset thereof) has established as the most significant areas of scientific inquiry.

Table 5 lists the percentages of scientists in each academic world who have published a majority of their work in major journals. Table 5 tells us about the scholarly norms that operate in worlds of science. Elites publish in the major journals more than do pluralists and communitarians. The differences are especially pronounced among communitarians, where only 26 percent of the scientists had published 80 or more percent of their work in the major journals. If we use a less stringent threshold and ask what percentage of scientists publish 60 or more percent of their work in major journals, the percentage rises to 63 for communitarians, but the percentage also rises to higher levels among elites and pluralists. Nearly all elites publish 60 or more percent of their work in major journals. Thus from tables 3 and 5, we can conclude that in moving from elites to communitarians, conformity in career practices decreases while the heterogeneity of commitment to and identification with scientific inquiry increases.

Acting an Age

The elite informant quoted above spoke of particular markers or milestones that symbolically place scientists at certain phases of a career. These phases include (but are not necessarily limited to) getting tenure, not resting on your laurels, keeping your "copybook clean," winning election to academies, sitting on national panels, and holding offices, in addition to the more formally organized rituals of promotion—the rise through clearly delineated academic ranks—which all compose a sequence of how the career, and the person experiencing the career, change over time.

These markers, and all others like them, are professional age-norms.

Table 5
Percentages of Scientists Who Published a Majority of Their Work
in Major Physics Journals

	Academic World		
	Elite	Pluralist	Communitarian
≥80% of published work	77%	71%	26%
≥70% of published work	86%	82%	32%
≥60% of published work	95%	88%	63%

Note: The journals considered to be in the category of "major physics journals" are those to which the physics community assigns greatest value. These include, in alphabetical order, *Astronomical Journal, Astrophysical Journal* (including *Supplement Series*), *Astrophysical Letters and Communications, Europhysics Letters, Geophysical Research Letters, Icarus, International Journal of Modern Physics* (including series A, B, C, D, E), *Journal de Physique* (including series I, II, III but excluding IV), *Journal de Physique Lettres* (now incorporated with *Europhysics Letters*), *Journal of Chemical Physics, Journal of Geophysical Research* (including series A, B, C, D, E), *Journal of Mathematical Physics, Journal of Physics* (including series A, B, C, D), *Lettere al Nuovo Cimento* (now incorporated with *Europhysics Letters*), *Nature, Nuclear Physics* (including series A, B), *Physical Review* (including series A, B, C, D, E, L), *Physics Letters* (including series A, B), *Physics of Fluids, Review of Modern Physics, Science,* and *Solid State Communications.*

They are turning points that define phases of the professional self. Bernice Neugarten's early theorizing and research on adult socialization explicates the normative underpinnings of the life course (Neugarten, Moore, and Lowe 1965; also Chudacoff 1989). Age-norms, a shorthand for socially defined rules that govern which events are to occur in the life course and at what times, function as a prescriptive timetable for the ordering of life events (cf. Roth 1963). "For a great variety of behaviors there is a span of years within which the occurrence of a given behavior is regarded as appropriate. When the behavior occurs outside that span of years, it is regarded as inappropriate and is negatively sanctioned" (Neugarten, Moore, and Lowe 1965, 716). As seen by elites, the professional life course is incremental. Individuals feel pushes and pulls from the larger world to perform according to age-graded expectations.[18]

The importance drawn from this normative, institutional view of the life course consists in the manner in which age-norms and age expectations operate as mechanisms of social control. At base, social control is a central sociological idea that explains how groups achieve, restore, and maintain order over time (Durkheim [1897] 1951; Janowitz 1975). In its concern with how order comes about, social control explicitly rejects the notion that social order—whether of a small group or of society at large— is a given; instead, it treats social order as a problem and asks how groups actively and routinely achieve this order through interaction.

Morris Janowitz (1975) has commented that many people mistakenly think of social control in terms of repression, such as surveillance and enforced conformity. The very idea of "control" seemingly contradicts democratic principles. But social control should be viewed in a more general sense, as "the capacity of a social group to regulate itself according to desired principles and values, and hence to make norms and rules effective" (Sampson and Laub 1993, 18; also Janowitz 1975). In this way, social control is carried out not only by formal means (codified sanction manifest in policies, bureaucratic codes, and laws) but also by informal means (sanctions manifest in face-to-face social interaction). Informal social control emerges from "role reciprocities" and interpersonal ties that bond members of groups to one another and to social institutions (Parsons 1951; Sampson and Laub 1993).

When reciprocities and ties break down, so does social order, as seen in any prison riot, dysfunctional family, or nation of unrest. Social systems cease to function effectively. The group in question, or society at large, must then restore order by introducing methods that effectively bring people under control. Social control thus is used analytically to understand a wide array of phenomena. It explains the underlying social basis of suicide (Durkheim [1897] 1951), just as it accounts for success in school (Willis 1977) and even for effectively run restaurants (Fine 1996). In this instance, it applies to how groups regulate passage through time by prescribing performance and behavioral standards to particular age-grades.

The idea of an age-graded life course should not be construed as an inflexible or monolithic series of stages through which all people inevitably pass. It is, rather, a general conception and socially desired unfolding of loosely defined, but key, phases of a career. Some people do not reach certain phases. Other phases may be skipped altogether, and the length of a phase varies from person to person, without sharp or clear-cut distinctions of where one phase ends and another begins.

I have argued that institutionalized careers in science are oriented mostly toward research. One way to again discuss the differences in the moral orders of each of the worlds is to uncover the relative frequency with which scientists achieve some of the symbolic distinctions that characterize their careers.[19] Major symbolic distinctions of science careers are presented in table 6, along with the numbers of scientists who wear these occupational merit badges.

The distribution of distinctions across the three academic worlds is significantly stratified. In each category, elites outnumber pluralists and communitarians by wide margins. The high-visibility fellowships most commonly awarded in the early phases of science careers are exceptions. These include the National Science Foundation Young Investigator Award (previously called the Presidential Young Investigator Award)

Table 6
Scientists Receiving Symbolic Distinctions in Science

Academic World	NSF Young Investigator; Sloan Foundation, Packard, and/or Guggenheim Fellowship	NAS or AAS Membership[a]	Elected Professional Office[b]	National Committee or Panel Member	Journal Editor[c]	Teaching Award[d]
Elite	11	11	3	7	4	6
Pluralist	8	0	0	0	1	2
Communitarian	0	0	0	1	1	1

[a]NAS refers to elected membership to the National Academy of Sciences; AAS refers to the American Academy of Arts and Sciences.

[b]"Elected professional office" includes such positions as president of the American Physical Society or officer (e.g., secretary, treasurer, vice-president, chair) of a professional society or board.

[c]Included in this category are scientists who served as editor, coeditor, associate editor, or guest editor of a professional physics journal.

[d]Scientists tabulated under "teaching award" are those who have won or been nominated for a university, college or school, or department award for distinction in teaching.

and the Sloan Foundation and Packard Fellowships, all of which are normally awarded before a scientist earns tenure. Also included in this group are the Guggenheim Fellowships, which are normally awarded to young senior researchers and scholars.

Pluralists come close to elites in number of award beneficiaries. This again may be due to the same two instrumental factors noted earlier: a contracted academic labor market that is pushing research-active and research-talented individuals toward a greater range of institutions; and a relatively nascent focus on research at once predominantly teaching-oriented institutions.

The other categories, like the major fellowships, are age-graded (with the exception of teaching awards). Elected membership to the National Academy of Sciences or the American Academy of Arts and Sciences— like election to offices, service on national committees and panels, and service as journal editors—typically occurs throughout a span covering the middle to late career.

The importance of these distinctions lies in their symbolism for the individual who is exposed (or fails to be exposed) to them. They set conditions for, indeed establish a whole milieu or culture of, the motivation to achieve. One could argue that cross-world differences in achievement stem from variation in individual abilities. But how is "ability" defined, as James Coleman has questioned, except as "the capacity to achieve under some standard conditions?" (Coleman 1969, v). A more profitable path toward discovering the determinants of achievement extends outside the individual and into the social environment that sets those "standard conditions."[20]

William Goode (1978) has offered a perceptive examination of prestige as a system of social control. Prestige is the positive response toward an individual or group for performances or qualities that are deemed above average. Represented by the many awards, prizes, honors, promotions, and citations, symbolic distinctions that mark the life course may all be seen as devices used by a group to inspire and regulate, not just to reward, performance. As Goode explains, "The acquisition, accumulation, expenditure, and loss of esteem or respect, and how [prestige or esteem is granted and withdrawn] controls the actions of both individuals and groups" (Goode 1978, 15). Where symbolic distinctions are few, control over individual performance, values, and attitudes is weak. Where there are many such distinctions, control is firm.[21]

Social control, then, becomes a primary way of differentiating life courses. Careers differ in part because individuals, by virtue of where they work, are exposed to different kinds and quantities of control mechanisms. As we have seen, environments themselves are set apart by their kind and quantity of control. Furthermore, individuals vary in their responsiveness to attempts at control: some comply, but others deviate. Consequently, careers in science, like those in any other field, assume various patterns.

Finally, in teaching awards (see table 6), elites again outnumber pluralists and communitarians. This finding might seem contradictory. After all, the public state institutions that make up the pluralist and communitarian worlds traditionally have vested their organizational identities in the seriousness with which they take, and in the quality with which they perform, the teaching role. This finding, however, probably says less about the quality of teaching in pluralist and communitarian worlds than it does about the self-consciousness among elites in recognizing achievement: theirs is a world singularly centered on striving, and they go to great lengths to mark phases of their progression.

Boundary Work and Identity

The depiction of a place vis-à-vis its similarities to and differences from other places constitutes a form of boundary work. Boundary work is a process in which individuals or groups actively construct an identity from the distinctions they draw between themselves and others who take part in similar activities. Part of this process involves drawing distinctions in ways that call attention to the more positive or flattering characteristics that individuals attribute to themselves or to the groups to which they belong. By doing so, they depreciate any stigmas that may otherwise be associated with their public identity. Boundary work also acts as a mechanism for groups to exert power over others by highlighting characteristics that uniquely define their position. Though this has been thought to be especially indicative of elites (Bottomore 1964), it is applicable as well to other groups, including those that make status claims in opposition to elites (for example, racial minorities [Gordon 1975]).[22]

Pluralists create a moral order through language that testifies to a middle position. What is especially noteworthy in how the scientists here described their institutions, however, is the way in which they constructed their middle standing as an advantage.

> The fact is that if you're at [this university] then it's going to be assumed you're second-rate, because this isn't one of the top ten departments. I think we're ranked [outside the top twenty] by *U.S. News & World Report*. But that's okay. That hasn't really bothered me that much. In terms of what I would consider a constraint, no, not at all. If anything, I've had far more opportunity here than I would have had, certainly more than a place like Harvard or M.I.T. If I had been offered a chance to go to Harvard or M.I.T. or Cal Tech over this place, I would have been flattered, loved it. My ego would have loved it. But there is no way I would have gone. I wouldn't have gotten tenure.[23]

Because the thresholds of performance are seen as lower than those for elites, yet not so low as to consider the institution out of the running for major scientific discovery, these scientists view their world as a happy medium. Many of them believe that the same scientific outcomes are achieved in their world without having to experience the onus of expectation that they believe characterizes the academic elite.

> [This university] has probably allowed me to mature more quickly than I would have if I had gone to Princeton or the University of Chicago. I think at Princeton or the University of Chicago they expect more in terms of prolific noteworthy work from their young faculty and I think I would have felt more pressure. I think I would have done exactly the same thing, but I probably would feel like I was about ten years older, because there is more pressure to perform. I think had I done what I did here at one of those universities, I still would have achieved the same outcome, it's just that I didn't feel that I had to do it here to be a success.[24]

Pluralists often juxtapose the conditions of their work with what they perceive to be the inhospitable conditions of more publicly visible and prestigious institutions.

> In terms of rankings of universities, you may have seen *U.S. News & World Report*, and this one, according to their list, for whatever it is worth, is somewhere around 20 or 25. But the top ones, they put Cal Tech . . . Harvard, Princeton, places like that. So it's considered very prestigious to be at one of those places. On the other hand, if you go there, most people are miserable, because there is so much pressure to compete, and so many egos trying to show off that it can be unpleasant. So from my point of view, I just didn't like that. You asked about roads not taken and one just came to me that I didn't even really consider, is that there was a job opening at M.I.T. The search was two years ago and many people came to me saying that I was the obvious choice for that job, given the parameters that were put onto it. I considered it very briefly and decided that even if I were to apply and was successful, I don't think I would want to be in that department permanently. There are just too many egos, too many turf battles. At a department like M.I.T., there are many distinguished people and many of them are prima donnas. They assume everything should go their way. They can be difficult to deal with. They also can be difficult at seminars. They each take turns trying to destroy the speaker. They can have a very narrow view on who is worth reading and things like that. From my point of view, I decided that just in terms of living I didn't want to be in that sort of group.[25]

Similar sentiments were conveyed by communitarians. As the physicist quoted below illustrates, elites (and some pluralists) are seen as impediments to, rather than facilitators of, important work.

> Looking far ahead, even as a graduate student, I knew that to have the freedom to do what I wanted to do and not have somebody pressure me to do something else, I had to go to a school that was emerging so to speak, where a young assistant professor could in fact establish a research program of his or her own making and not do what the senior faculty wanted them to do.
> *Did you consider other places?*
> Oh yes, I told you. I did. I knew that it had to be a relatively small school. Preferably state-supported. Preferably in the South. I didn't want to get into the kind of rat race which you get into in some of the more prestigious universities.

Why?

Again, you're under too much pressure to go down the roads other people believe are significant and you've got to get research funding or you don't get tenure and you don't get retained.[26]

Another communitarian agreed:

I definitely would not have liked an environment of "cut-throat" and where people are miserable, where people are always unhappy because it's how it goes. I think there are certain types of universities. I think M.I.T. is that way. I guess all the first-grade or first-order universities in general tend to be that way. And maybe this is why I didn't go as a graduate student [to these types of institutions]. I applied to some universities, like Yale and Princeton, and got accepted, quite a few, but in the end I really preferred to be in an environment where you can work and study, but also be happy and not necessarily have to be miserable . . . I like a friendly environment. I like to love to come to work. I've been with people who have gone to these universities and these people are dishonest, they are cut-throat, they will do anything to get something, their value system is totally different from mine. They are a different breed the way I look at it. I didn't want to be one of those people. If you are in an environment like that, you may have to become like that.[27]

The manner in which pluralists and communitarians describe elites is altogether different from how elites describe themselves. Our initial informant described the people with whom he works, and the environment in which they do science, as prized resources. The reality of the elite world is constructed differently by members and nonmembers. For members, it is to one's intellectual and scientific advantage to work in the presence of those who may be taken by nonmembers to be overly competitive and egoistical. An elite put these distinctions in context:

There are two simple reasons you want to be at a place like [this one]. One is that your colleagues are good and you can talk . . . when I go to ask my colleagues questions I get good answers, because my colleagues are knowledgeable. And then you have good students and you have good post-docs, so it's only to my benefit to be at a good school, because I'm going to benefit from the people around me. And it also means that there's a steady stream of good speakers coming through the area who want to come here and talk about their ideas. I would have to say that I've had every opportunity, given my position, to do whatever I wanted, because I'm at a good place, with good people around me, with good speakers coming through. I think if I was at the University of Kansas I would be higher on the pecking order. Probably at most other first-rate physics schools I would be in the middle. I'm not interested in being at the top. I'd much rather have colleagues who know more than I do.[28]

Although pluralists see advantages in their position, they realize it is not without its costs. They deal with stigma (or at least a comparative lack of prestige), in part by the ways that have been described in the accounts above. They also deal with a set of structural constraints. Pluralists often raised the subject of students.

> My biggest complaint I guess is the lack of good students. I have a couple of students. One of them will probably do okay, and the other will be a real struggle. In contrast to other places where you can give a student a problem and they will just take off and work with it, I basically feel that I have to explain many times until they get it as opposed to just mentioning to somebody, "oh you should read about this" and having them come back and explain it to me. It slows things down. My game plan was to put this [part of a project] in a student's lap, and I estimated how long it would have taken me as a graduate student and doubled it. It turns out that doubling it is not enough. It's more like a factor of six or something.[29]

> Of course you want to go to the Ivy League so that you have better people, better students. Ivy League is better, but that's alright. We will do the best here.[30]

> The main disadvantage of being at an obscure university like this one is that the students are on the whole not as good. For instance at the Ecole Normale in Paris, where you have highly selected students, about one in a thousand, you notice the difference. They are extremely quick. They are very challenging to work with. The sort of student you get here is not so quick and not so stimulating. That's I think the key difference. It's the quality of students.[31]

Conclusion

The groups that make up the three academic worlds of science display distinct moral orders that orient and guide ways of knowing and behaving. I have described these moral orders on the basis of commitment to and identification with science. We have seen how the worlds fall into a spectrum that widens as we move out from elites to communitarians. In looking out over this distance, we see progressive changes. People become more heterogeneous in their commitments and self-identities. They become increasingly tolerant of different careers, if only because variety is the norm of their world. These differences are more broadly conceived in terms of expectations. Each group creates its own meaning of achievement, extraordinary achievement, and lack of achievement. Success and failure in one world does not directly correspond to the same success or the same failure in another.

Worlds differ on an additional count. Only in one, that of the academic elite, do we find people who clearly articulate a uniform moral career. The career and the person undergo successive changes; they bear different images at different phases. It is a world in which success is seen in terms of vertical mobility. Opportunity to get ahead defines the ways people involve themselves in work. Symbols of achievement—medals, honors, and esteemed people themselves—serve as constant reminders of an individual's purpose in belonging to this world.

The same is only partially true in pluralist and communitarian worlds, which contain multiple and competing systems of reward. The career, nowhere near as institutionalized as it is among elites, takes on varieties of forms. Phases of the career and the person are only nebulously conceived. Where careers revolve around teaching, the life course is relatively unstaged.[32] As for elites, opportunities for making major gains in status lie in adopting a research career or in embracing full-time administration.

In outlining the moral orders of the three worlds, I have described what it is like to belong to them. This mapping of the moral orders thus provides a basis on which to begin a discussion of how these sixty careers have unfolded in context.

THREE

Professional Pasts

This chapter will trace the professional paths that have shaped how scientists know themselves in the present. I will look at how the scientists viewed themselves and their futures just before beginning their careers and how their visions evolved to the points where we find them today. Accordingly, the scientists take us (and themselves) back in time, normally to the end of graduate school, when the idea of and the commitment to a career in physics began to crystallize at a resolute level (Becker 1960; Becker and Carper 1956a). Each of the scientists underwent a process of professional socialization during this period. All were the targets (and at times the victims) of rituals, of rites of passage: the examinations, the experiments, the theses, and the public presentation of work and of self before an adjudicating other. Their lay identities developed into professional identities.[1]

I asked the scientists to describe their aspirations: what they wanted, or hoped, to attain. They provided pictures, sometimes hazy but other times crystal-clear, of the future as they saw it, and they indicated some understanding of where they saw *themselves*—their role, their stature, their anticipated identity—in the pictures they drew.

I have a theoretic objective in raising the topic of how scientists account for their pasts. This objective concerns how scientists conceive and act on ambition, if they do so at all. I seek to identify how the scientists evaluate their progress in relation to their initial achievement expectations. I will thus present a chronology of ambition (cf. Chinoy 1955, esp. chapters 9 and 10). I will focus on how ambition gets (or fails to get) constructed and reconstructed, and on how ambition changes—thriving, dying, and stabilizing—differently for different groups over time.

I will set my discussion in the context of how scientists conceive of the moral career because the moral career defines what all scientists, at some point, construe as an expected passage. In principle and often in practice, scientists form and evaluate self-images using the moral career as a backdrop. It provides a script and timetable for how careers unfold and at what rate, but of course not all or even many scientists adhere to the course it prescribes (Buchmann 1989; Schank and Abelson 1977; White 1993). Although this template is moral, it is not modal. People living out careers in science—like people living out other careers—conform or deviate from the template in different degrees. In hearing of the paths that people have traveled, we will see how some individuals' careers have closely followed the moral career whereas other careers have represented twists, meldings, breakages, or outright rejections of the template. In other words, the careers of these sixty scientists are not uniform. They fan out in many directions, at many different speeds.

I will also postulate that elites, pluralists, and communitarians gravitate toward three distinct and central career tendencies, or typical pathways, as their careers unfold. After describing these major paths, I will examine how each of the three groups differs internally. I will suggest that though each is characterized by a typical passage, each also contains individuals who deviate from the dominant pattern. These individuals often exhibit characteristics similar to those found in adjacent worlds. This suggests that although academic worlds are distinct on several dimensions, they also seep over into one another.

Looking Back on Beginning

The Quest for Recognition

All of these scientists began their careers with expectations about how those careers would proceed. The expectations ran a gamut—from the most basic to the most grand, from the most loosely articulated to the most certain. This variety was met, though, with a conditioned uniformity rooted in what these scientists all perform: the scientific role. All of them expressed the institutionally mandated desire to involve themselves in "good," "interesting," or "important" work.

> [I wanted to] just do good physics and enjoy, make some contribution—obviously to enjoy what I was doing. It's like doing something that you are very interested in, and the pursuit of that is something which is both hedonistic and perhaps does something good for the field.[2]

> I was very interested in research and doing experiments, in the thrill of discovery, so to speak. And that has been very satisfying. At some level every physicist would like to win the Nobel Prize, but I certainly

realize that takes a lot of luck as well as skill, so I don't think people go into it with just that as a goal. I think you want to have a satisfying career and have fun doing science.[3]

The presumption is that by carefully selecting "important" problems and following through with "good" work, scientists may carve a path toward major scientific breakthroughs. They seek to unravel long-perplexing mysteries of nature and the structure of the physical universe.

Such aspirations in part reflect the scientists' genuine wishes. Ideas, abstract reasoning, and intellectual challenge play a leading role in drawing people to science. This academic orientation also frames the hopes on which their careers would center over the long term.

But these outlooks also reflect the way in which scientists view themselves as rational, objective, and dispassionate members of a scholarly community. Admissions of "aspiration" or "ambition," even if involving desires that rely on ethical means toward a scientifically legitimate end, are met with suspicion by the science community. Part of the responsibility of playing the scientific role involves pledging allegiance to its institutional purpose: the development and extension of certified knowledge. By expressing the prosaic, institutional goal that all members of the science community adopt or are supposed to adopt—"doing good work" or taking a disinterested approach to scientific inquiry—individuals deflect attention away from personal ambitions and toward the sanctity of the scientific role.[4]

Apart from discussing aspirations connected to their social identities as scientists, the physicists discussed sides of themselves that are normally kept private. They spoke on a more personal level about how they hoped their careers would turn out. Recognition plays a vital part in how science careers are shaped and in how they are subjectively experienced. For many, the enduring quest for recognition explains the birth and life of an overriding ambition.

Why is recognition so important? The answer has been explored extensively by Robert Merton (1973c) and is quite simple: recognition from people who are competent to judge a contribution is the prime indicator that a scientist's work has contributed to knowledge. Recognition, in its many forms, conveys impact, in its many degrees. As Merton explains, the quest for recognition is thus institutionalized. The quest is compatible with, indeed satisfies, the goals of science, so long as in practice, scientists remain bound to ethical guidelines. This is conveyed in the apt line, "Recognition for originality becomes socially validated testimony that one has successfully lived up to the most exacting requirements of one's role as a scientist" (Merton 1973c, 293).

Following Merton further, we see that the quest for recognition is linked to concerns about priority in discovery, that is, to the establishment, by the science community, of who is the first to make a contribution and is thus accorded due reward. To an outsider, concern about priority may appear trifling and, worse, indicative of a needless compulsion. Priority, however, is significant and is seen as such by scientists: it establishes on principle who has contributed what and thereby identifies on principle who is to be rewarded. It thus certifies for the record what work is original and, in doing so, affirms the goal of science to extend certified knowledge. Put differently, priority operates as part of a functional system that gives credit when and to whom credit is due.[5]

Why scientists seek recognition and what they derive from it, however, consists of more than only a realization that their work advances knowledge. Recognition brings one closer to a place in history, and thus one's influence is felt by future generations. This is not to say that all scientists seek fame or glory but that all, or nearly all, scientists at one time or another in their careers seek to have an impact on the world that they study and in which they live. Recognition is a measure of one's self-worth and, to this end, is a basis on which identity is built. Great recognition, for great accomplishment, holds the possibility for a heroic identity, sowing the seeds for ambition. Yet even though recognition carries this intense personal importance, the importance itself is a manifestation of being a socialized member of science, for originality stands as the ultimate measure by which one's work is judged.

What is more, the accrual of recognition translates into subsequent scientific opportunity, constituting a kind of professional lifeline. The process and consequences of gaining and failing to gain recognition have been described as the accumulation of advantages and disadvantages (Zuckerman 1989). Small differences in the early stages of careers amplify themselves in subsequent stages and result in disparate career outcomes. The "Matthew effect" (Merton 1973a) is a special case of the more general mechanism by which recognition is bestowed. It states that recognized or eminent scientists have a propensity to receive disproportionately greater credit for their scientific contributions than lesser-known scientists, who receive disproportionately little. Recognition conveys the perceived significance and quality of a person's work. Its importance is thus realized even by scientists who are the most steadfast in upholding core scientific values.

School Background

Scientists' achievement expectations were clearly defined by their experiences in graduate school and immediately thereafter in postdoctoral positions. Socialization in this period (and over the entire life course) includes learning what groups expect of their members so that in principle, individuals will adopt those expectations as their own. Graduate school was a time when these individuals came in closest contact with, and could most resolutely anticipate leading, science careers. At this point, the scientists learned many of the informal lessons on what a scientific life is like and how science careers look, however imperfect or incomplete these lessons may have been.

Although the scientists now have careers in three distinct types of academic institutions, many began at similar starting points. The schools at which the scientists received their doctoral degrees are listed in table 7. In comparing the schools, we see that elites and pluralists are more similar to one another than communitarians are to either of the other two. Fourteen of twenty-three elites (61 percent) and twelve of eighteen pluralists (67 percent) earned their Ph.D.'s at institutions considered to have one of the top-ten departments of physics in the United States.[6] Only three of nineteen communitarians (16 percent) earned Ph.D.'s from a top-ten de-

Table 7
Graduate Institutions of Scientists, by Current Institutional Identity

Elites (N = 23)	Pluralists (N = 18)	Communitarians (N = 19)
Berkeley	Berkeley	Arkansas
Birmingham	Chicago	Berkeley (2)
Bombay	Columbia	Cal Tech
Cal Tech (2)	Cornell	Colorado
Chicago (2)	Harvard (3)	CUNY (2)
Cornell	Illinois (3)	Georgia
Harvard (2)	Indiana	Iowa State
Landau Institute[a]	Johns Hopkins	Louisiana State
London	Maryland	Minnesota
Milan	M.I.T. (2)	Missouri
M.I.T. (5)	Oxford	Nebraska
Minnesota	Pennsylvania	NYU
Northwestern	Purdue	Syracuse
Pennsylvania		Tokyo
Princeton		Wisconsin
William & Mary		Wroclaw[b]
		Virginia

[a]Russia
[b]Poland

partment. These schools should be kept in mind because they will bear on my later discussion of how the achievement expectations of scientists are framed and how those expectations evolve over time.

Imaginary Careers and Imaginary Selves

Ambition, and visions of a career, were often modeled around the attributes of others with whom scientists had contact. Some of these were face-to-face contacts, often with the scientists' mentors. Other contacts involved intellectual forebears who, by having worked in the field, functioned as exemplars around which newcomers could claim status by association (cf. Foote 1951). These forebears were the time-honored heroes of the discipline: Copernicus, Einstein, Keppler, Newton, Ptolemy, and many others. All of these contacts—living and dead, down the corridor or in the physics pantheon—stand as icons of an ultimate vision. They represent that which is desired, serving as examples that scientists want to follow (if only when young) because, in Weber's terms, these figures are imbued by the graces of charismatic authority. They inspire interest, fascination, and emulation. Through the eyes of these young scientists, graduate school was thus a proverbial age of expectation, if not of innocence.

> I wanted to have my advisor's job. I wanted to be a professor at a university. I certainly was always shooting for the highest position. I just wanted to keep going up and up and up.[7]

In accounting for their passages through this momentous stage in a scientific life, the physicists' narratives take a dramatic form. The scientist is alive with hope. Young life embodies promise and potential. Confidence has been built. The individual is strong and unyielding, a center of activity. Energy is wanting to be spent. The exact future may seem uncertain, but these individuals know that they will play a prominent part in it. The world awaits them.

> My aspirations were to work with Feynman[8] for the rest of my life and to be like Feynman, because he had been my Ph.D. advisor until he left Cornell. He left Cornell, and then I switched to Bethe.[9] But basically I was discovering how much fun it was to roam around and think about things with total freedom. I wanted to do good work and be rewarded for it by friendships and by being invited to meetings and by being treated like somebody who knew and who could teach. I was looking for respect.[10]

The scientists' reflections on their early aspirations reveal that a world of illusions animated how they envisioned the future and the role they hoped to play in it. George Herbert Mead advanced the idea of "taking the attitude or role of the other" to describe how individuals engage in complex patterns of behavior typically exhibited by the "other's" social position (Mead 1934). He used these terms to explain the conduct and social development of people who had acquired at least the elemental characteristics of those whose attitude or role they adopted (Cook 1993). The implication of this idea is that socialization, in youth as well as throughout the entire adult course of life, is based on imitation. People conduct themselves according to patterns they observe in others—but of course not *all* others. Attitude and role-taking is anticipatory: individuals model their conduct on behavior patterns associated with the positions they anticipate holding. Thus objective status (who one is) and subjective status (who one thinks he or she is or ought to be) can be radically disparate. People pretend to be the person they anticipate becoming.

This tendency is profoundly evident in the narratives of the would-be scientists. An individual plays the role of graduate student but takes the role of demigod. A prominent feature of passing through phases, before the actual onset of the academic career, consists of creating scenarios in which the individuals cast themselves as heroes performing magnificent feats, scaling unknown scientific summits and winning the respect and admiration of others.

> When I first went to Cal Tech the two most powerful influences there on anybody were Feynman and Gell-Mann,[11] so the obvious first thing that I think everybody including me at that place hopes to attain is something at that level. That would have been a naive first hope. One realizes pretty soon that that is not possible. After interactions with them you realize it could be very, very difficult. I think there were two aspects of it. One was the extraordinary originality. They were looking at things in ways that you hadn't thought about and presumably you could have had you been that clever. The other is of course speed. They were very quick at doing problems that took you a longer time.[12]

> I used to think I was going to win a Nobel Prize in physics. That is something I no longer obsess about very much. I think that's sort of the general thing that physics graduate students think they are going to do, maybe not a Nobel Prize, but something very fundamental. I was at Cal Tech when Feynman got his Nobel Prize. Just when I came there Gell-Mann got his Nobel Prize; it was a very exciting time there

because of those things. Of course, Cal Tech is sort of the pinnacle of the world in terms of physics. I wanted to make my mark in physics in some spectacular way, but I think that is just youth. Although I think I'm good at a lot of things, I don't think I'm particularly of that class to do that.[13]

Ambition differed by the type of school where the scientists completed their doctoral studies. The impact of place on the person is again felt. Visions of greatness were normally found in the early aspirations of scientists who had elite graduate school backgrounds. Notice how the physicists above refer to the sense of glory permeating the environment during their student careers at Cal Tech. Their sense of self was framed by the achievements of those with whom they had close contact. Symbolically, they were the direct beneficiaries of an enormously potent and powerful scientific estate.

In being part of the elite world, one learns a *cultural script* for success. Members are socialized to expect high achievement from themselves and from others because great achievement marks a moral passage. The cultural script for how to act out a career represents the major channel through which social worlds transmit norms to individuals.

I had been trained to expect a certain class of jobs and also been trained that if I did not attain that class of jobs, that that was unacceptable or to be looked down upon. As you go through graduate school you are trained by your mentor, your advisor. He has a certain type of job, which is research; he has independence. He decides what he wants to do. You also see who gets respect and who doesn't at professional meetings. You get the impression that people who wind up going to more applied areas, more development areas, are apologetic about it when they come back and talk [to people at] the university where they got their training.[14]

By contrast, schools with more modest reputations instill in students a more modest, and often a more pragmatic, vision of their place in the future. In chameleon-like ways, individuals assume many of the characteristics of the institutions in which they were socialized. We see this pattern not only in the scientists above, where success is the norm, but also in those whose school backgrounds make success seem more like an exception.

A scientist who completed his doctoral studies at the College of William and Mary is a fine case in point. Given his school background, a label of limited potential was applied to his life chances by the wider society (but not necessarily by himself or by close associates). Such labels are applied based on a socially held assumption that "success" is stratified in a system of vertical classification.[15] Particular types of schools are thought to lead to certain levels of achievement. Graduates are labeled in ways that estimate the perceived value and potential assigned to the school that they attended. This process often occurs independent of knowledge about individual abilities. Thus, graduates of elite institutions are "promising," even though some are academically deficient. And graduates of other institutions may be discounted, or may be overlooked altogether, even though many possess notable talent.

Because the scientist from William and Mary now finds himself in an unlikely position as a member of the academic elite, the genre of his biographical account takes the form of success beyond wildest expectation. He was not expected, nor did he himself expect, to realize such achievement.

> I went to Wittenburg University [for undergraduate college], a little liberal arts college in Ohio. It was quite nice, but in terms of a physics background, it was about half of what we give the physics undergrads here. It just wasn't a graduate student training prep school. Then I went to the College of William and Mary as a graduate student. That was again the best I could do. My funny joke that I don't tell too many people is that I didn't get in here as a graduate student. Now I have tenure here. I'm a person that runs on fear. I wanted to make sure I didn't screw up. I didn't want to teach badly at this highly ranked school which I had already thought I couldn't even get into. This is a spectacular place. It has boosted my career way beyond what I ever could have imagined.[16]

Ambition for greatness before the onset of an academic career is not a customary feature of communitarians. Recall that they more often completed their graduate training at pluralist, rather than elite, schools. Compared with elites and pluralists, communitarians were more limited, and perchance more pragmatic, in the scope of their early visions. Like the William and Mary graduate, these individuals call attention to the pattern of espousing modest visions that correspond to the more modest visions of the places where they went to school.

> I had no goal. I didn't have a goal to do this or that or the other thing. I was not very ambitious. I had a lot of other interests besides physics, and the fact that I devoted so much of my time to physics was not something I would have expected. I was athletic; I spent a lot of time

playing tennis. I did other sports. When I was younger I spent a lot of time with lots of different women. If you would have asked me when I was a young man if I was going to be a physics professor when I was sixty years old, I would have said probably not.[17]

I thought I wanted to teach at a university and do research and not much else as far as professional aspirations are concerned. I hoped to eventually become a tenured full professor, but I didn't have any grandiose expectations or aspirations. I thought most of my peers wanted to do more or less what I wanted to do—that is, get a university post and combine teaching with research. I think some of them inclined towards teaching and wanted to make that their main focus.[18]

The Fog of Youth

The three academic worlds that scientists entered as budding physicists presented special problems for how they would act on their ambition. The divide between one's expectations and the capacity to realize those expectations within the limits of ability and environment creates a personal crisis, the resolution of which consists in a coming of age. "Achieving maturity" describes the way the past looks in the present.

A graduate student is really a baby. A graduate student has very good intellectual knowledge of the big picture, but a graduate student is a baby in terms of the long haul through life, the many day-to-day, leap-to-leap problems that will occur. In a sense you're so busy as a graduate student that you don't have time to develop any real perspective on what life is like. The changes that occur are that you can look at yourself—this is one of the agreeable aspects of aging—and understand yourself and your career in ways that you can never do in graduate school, because you're too close to it, too enmeshed in the whole thing. When you're young, you shoot very high, but you also kind of shoot from the hip. As you get older, you learn more about yourself and your place and so you adjust, you take a more accurate shot.[19]

This scientist depicts the unfolding of self-identity in the third person. He accounts for his passage but generalizes about its applicability. A developmental theme of these accounts emphasizes an evolution from youthful idealism to realistic self-appraisals.

My choice of problems is very programmatic. When I used to write my grant applications, as a young theorist, what I did and what I said I would do had a very uncertain connection. Whereas now when I write my grant applications I pretty much stake out problems that I know I will make progress at. The aspirations are in a sense longer-term. Younger theorists sniff the flowers and try to make a quick killing. Whereas now, towards the last 25 percent of my career, I want to have projects that I know I can make a contribution to and I know will have some impact. So it's a much more calculated time in my life than when I was young. The aspirations change in that you have a better understanding of who you are and how you fit into the professional world, and you guide your actions accordingly.[20]

When you are twenty-five you think you can find the answer to the universe, and when you are fifty-five you realize you are going to be dead in twenty years and you're not going to find the answer to the universe and it changes your priorities. There is a lot of romanticism in physics, but I think there's a quote from Steven Weinberg[21] where he says, in the end, the more we understand the universe the more pointless it seems. I think as a young man I thought of physics as something that would yield ultimate answers in sort of a religious sense. But I don't feel that anymore. That takes some of the intensity off of it.[22]

The beginning of a career puts in motion an evolution of self. Starting out often appears youthful in hindsight and is understood through a number of associations with youth: idealism, romanticism, fearlessness, immortality.

The evolution in self-views stems from the experiences that people encounter in their working lives. Self-appraisals of ability and of chances of making a major discovery are often based on an ever-clearer realization of the type of talent (and luck) needed to accomplish such objectives.

You get to understand how hard it is. When you are a grad student you fantasize about following in Fermi's[23] footsteps. Then you realize—it takes a long time to realize—how smart they really were because they make it look so easy. All these great men make it look very easy. You can imagine doing it. But that's because they did it. Everyone is romantic when they are in graduate school. I thought I would be another Fermi. It was a romantic view. I realized how smart Fermi really was, after a career of trying to do original work. He was an unusual individual. This is somebody who lives once a century.[24]

As a graduate student, I was quite ignorant about what problems can be done and what cannot be done. I had aspired to solve problems which required more brain[power]—impossible to solve—things about the original universe, how the universe evolved at the very earliest time. And now I know better. Now I know they are unrealistic goals. We don't have enough understanding of physics to be able to address these questions at this point in time. So my goals are more realistic now. As you learn more science you realize there are more realistic goals. The answer is rather simple: as you learn more, you find out what is possible and what is not possible; what is something that can be done, what is doable and what is not doable.[25]

The changes in self-identity that occur in light of experience constitute a dynamic aspect of the moral order in which people work. Changes center on a fundamental question: how do members of an occupation identify with their work in ways that take into account issues of blunted mobility and failure? This question has been the subject of some previous sociological inquiry, resulting in generalizations that emphasize the central developmental theme. "A predominant theme in the literature on socialization presents life as movement from youthful idealism to more or less realistic mobility motivations, more or less contented adjustments which progressively confine behavioral potentialities and imputed identities within an acceptable range and to which an individual becomes committed" (Faulkner 1974, 136).

In a comparative study of orchestra musicians and hockey players, Robert Faulkner (1974) found that members of "feeder" or minor-league groups began to exhibit a set of behaviors after realizing their dim chances of ever making it in the majors. They stressed the personal benefits of their present position; they commented on the costs involved in advancing in the career hierarchy; they strived for average recognition rather than for outstanding accomplishment; they shifted their primary interests away from work to other areas of life, especially to family; and they stressed a calculative or utilitarian identification with work.

We find many of these patterns in the adjustments that the scientists made after entering their respective work worlds. For example, we saw in chapter 2 how pluralists stress the advantages of their location: they see themselves as conducting major scientific research without feeling the strain that they think afflicts elites. But in other respects, comparisons among scientists, musicians, and athletes are not parallel. Not all scientists aspire to careers at particular types of institutions, although most have at least vague ideas, if not well-honed ones, about the various types of academic settings. Careers in science are organized much less monolithically than careers in sports and the performing arts. In the latter groups, there is a clear-cut career line. Performance in the minors makes or breaks a career in the major leagues. Career expectations among minor-league players are thus uniformly directed at the goal of being selected by a major organization.

Jonathan Cole and Stephen Cole (1973) have asked how individuals in a social system such as science handle failure. Failure should of course be construed in comparative, rather than absolute, terms because few scientists (or others) see themselves as complete failures. Most are inclined to define themselves as successes in ways that their local environments allow. Hence we could speak just as easily of relative success. The Coles make the point that it is difficult for scientists to find reason for their "failure" in the structure of science because they believe science is equitable. Accordingly, the Coles claim that scientists have to turn to themselves to explain lack of success.

Whether or not scientists view science as equitable is, however, an empirical question.[26] It is also plausible that those who see science as equitable are scientists who have reaped the greatest rewards and are thus in a position to attribute their success to their own ability. Conversely, less recognized scientists may more readily see partial workings in the institution of science (while perhaps also seeing failings in their own ability) and thus may attribute their standing to both themselves and the structure of work.

It is also conceivable that differences among and within fields of science prevail. In times of failure, physicists may be more likely to blame their departments, on whom they depend for resources and support. Mathematicians, who are more often solo practitioners, may be more likely to blame themselves (Hargens 1975). Similarly, theorists may be more prone to blame themselves because they often work free of other major wants and needs. By contrast, experimentalists may be more prone to blame others.[27]

The adjustments that scientists make over the course of their careers are considerably context-specific. The expectations they bring to their work are numerous and diverse, rather than uniform. In addition, individuals' priorities vary in ways that reflect the multiple systems of reward in an academic life. The ways in which the self-identities of these sixty scientists began to unfold depended highly on the academic worlds in which they had studied and in which they subsequently worked.

Interworld Differences: Points of Professional Destination

Erving Goffman developed the idea of "cooling-out the mark" (Goffman 1952). "Cooling-out" is a process in which people adjust to situations that fall short of what they had been led to expect. In its enactment, "the cooler" substitutes alternative "prizes" for the prize or goal originally sought by "the mark" in order to make failure to win that original prize seem less disastrous. "The cooler has the job of handling persons who have been caught out on a limb—persons whose expectations and self-conceptions have been built up and then shattered. . . . [T]he disappointment of reasonable expectations, or even misguided ones, creates a need for consolation" (Goffman 1952, 452). Indeed, the process of cooling-out the individual can be accomplished so artfully that the substitution appears in the mark's eyes as the prize that he or she originally wanted—hence Goffman's use of con-game metaphors.

All of the scientists were cooled-out after embarking on their careers. Their expectations adapted to the situations in which they found themselves as young physicists. But their careers were cooled-out in different ways. To extend the metaphor, their careers cooled to and then proceeded from different points on a professional thermometer.[28] The worlds that they entered allowed their careers to proceed in many different directions, at many different speeds, and with many different sets of expectations. The results are seen in the adaptations that people make in their new environments. Foremost among these is the change in the view of the self. People reconsider, and often rescale, the aspirations that brought them to points in their early careers.

Where Goffman makes a social project out of cooling out the mark, here individuals do it for themselves. Each of these scientists establishes a self-definition from available opportunities—giving rise to a "situated self-identity" (Mills 1940). A major task is to bring about a match between their self-definitions and their available opportunities. As I have begun to explain, the process of reading one's situation to "find" one's self-identity is initiated in graduate school, when socialization and the formation of a professional self-image (however grandiose or self-deceiving) begin in earnest. Once this process has begun, it holds all along the life course. This process constitutes a way toward self-understanding because it takes the cumulative consequences of everyone's behavior and fashions the symptoms of an individual's self-identity.

Communitarians

The starkest changes in outlooks on self and career are seen in communitarians. The academic world that they entered presented sharp contrasts to the worlds that they had known as students. This new world was, for most, one they had never known.[29] Some had come from the elite. Many had come from pluralist schools, where the research programs often closely approximate, and at times are identical to, those of the elite. Like several other scientists, the one quoted below spent his graduate career at a major research institution with an illustrious history in physics, the University of California at Berkeley.

> I thought I could maybe make some nice discoveries, but I had realized early on that the probability of making a major discovery like Einstein or Newton, those kinds of major discoveries, would probably be completely out of my grasp, reality being what it is. But I thought maybe with some luck (a lot of interesting results come out of just pure luck), [I could], if I stumble around and find something that hasn't been observed before. I thought, well, there is always a reasonable chance that I might be able to find some new thing. I realize that's not going to be possible for me to do, to accomplish much more than I already have, considering the rate at which I've made progress. Over the time that I've got left to do productive work, the chances are I won't really make any major or even minor discoveries. I think that if one has limited talents, then you are going to make rather limited contributions. I think now that my contributions will be rather limited. I can look back and see what opportunities slipped through my fingers.[30]

Structural constraints imposed by the communitarian world were evident in how people viewed the reaches of their involvement and impact in the field. Constraints took many forms but were always thought to place limits on how far one could go. They included lack of equipment, small department size, a low density of expert colleagues in specialty areas, low graduate student enrollments, limited research assistance and funding, and less access to mainstream journals (see table 2). To speak of structural constraints, then, is to speak of what opportunity looks like for those on the inside.

> I've always been kind of a lone wolf–type person; I haven't had many collaborators. I've worked with students and post-docs, people like that, but I haven't really collaborated with the big dogs in the field. [If I had], obviously it would have made my contributions greater, because I would have been working with people that had more resources. I could have contributed more that way, simply by joining in a bigger effort. It would have probably enhanced my reputation because I would have been associated with people that [other] people would know more about. Just to be in the swim, so to speak, being associated with these other groups.[31]

I wish [this school] had a better reputation. I think it still has got the teacher school reputation. It would be easier to have gotten funding at Duke than it would be to get funding here, because Duke is a better-known school, as far as Washington, D.C., is concerned [where major science funding agencies are located].[32]

I have no collaboration. That's been the problem with my research. All the papers that I've published have been just me. That's difficult to do, because what I found, even though when I applied for grants they [reviewers] agreed that the problem looked interesting and relevant, I was still not able to get federal grants simply because I didn't have collaboration. And so without funding in this particular day and age, it's very difficult to find the time and to find the resources for doing research. That kind of died out.[33]

For some of the scientists, places are paralyzing. Expectations are completely dashed. Their professional life bears no resemblance to what they had earlier envisioned. What was previously viewed as professional work now seems more like a "job." In ways unique to communitarians, scientists were sometimes cooled-out to the point that they froze professionally. In neither of the other two worlds do we find an account of a career that has basically come to a halt. In such a career, only the bare requirements of the academic role—coming to class to teach—are met, and then only with some reservation. In the small departments of which the communitarian world is made, professional deaths are not secrets (nor would they be if they occurred in larger elite departments).

One [guy] looks upon his work here as sort of a job. He has a lot of outside interests and it's not much different than putting in your time as a bank manager. I think when he was younger he had hoped to have a career involving both teaching and research, and he found it difficult to continue his research. It was probably [because] he didn't have sufficient interest in his research in order to persevere in a department where you are very much on your own. We didn't have good machine-shop facilities; we didn't have technical help, and so you really had to do just about everything on your own, and it was not easy. If you were at, say, Harvard or Yale, you would have a lot of support; your teaching duties would be lighter. So it was extremely difficult. He just didn't want to put in the level of effort required. He was not promoted to full professor; he remained an associate professor.[34]

Even those who did not enter the communitarian world with an elite school background, or with the grand hopes that such a background normally inculcates, had to make adjustments in outlook. As pragmatic as some of their initial aspirations were, many communitarians found that their world could not satisfy their goals or interests.

> I went to Ohio State [as a visiting assistant professor] and became involved in the academic rat race to some extent. I wanted to go to a university like Ohio State or another Big Ten university or some university that had a pretty well established graduate program. I have to confess, I was a little disappointed after having done postdoctoral research in Europe and being at Northeastern University. I just felt that [this school] wasn't very well established in physics, and so I felt a little bit of disappointment.[35]

> I had a pretty good idea what I wanted to be doing twenty years down the road, ten years down the road. I was envisioning developing a relatively strong working group. If I could bring in people who could support me and what I was trying to do, and I could interact with them, we could have a pretty good working group in, say, astrophysics, stellar evolution, or in nuclear astrophysics. Initially it looked very promising, [but it never came to pass]. So my aspirations were blunted at a relatively early stage. Within the first five or six years I could see that nothing was going to really be fulfilled and promises were not going to be kept. I wanted to be doing research, I really wanted the research to work extremely well, as most researchers do. I would say that my aspirations haven't been fulfilled at all. To that degree it's been very unsatisfactory in my mind. It's been very unrewarding. My professional development has been somewhat thwarted. I'm not doing today what I had expected I would be doing. And so you learn to release those kinds of fantasies and deal with reality and take what's given to you. You make the very best you can out of it.[36]

There are, without question, happy and contented physicists in the communitarian world. But there are also scientists who sound bitter, as many of the accounts above convey. Those who are critical of the institutions in which they work are most likely to be found among communitarians. There is a culture of complaint, an airing of unhappy sentiment. This world seems to suffer the most political strife—at least its members are inclined to include tales of political unrest in their biographical narratives.

> The chancellor in his faculty convocation at the beginning of the semester said this is a school of "perpetual possibilities." When all of the

faculty laughed, I think he really didn't understand what he just said. The possibilities *have* existed here, it's just that that's not enough: you have to actually fulfill some of those. There is a chance to have a material science program here. The only opposition seems to be coming internally, and if you ask me why—I don't know. I think the university is being run as a business. The problem [here] is that it's always the best people that leave because they are the only ones that can. So what you end up with is the worst part of the faculty left over.[37]

When it comes to higher education, what the politicians [in this state] say and what they do is the difference between night and day. They are misleading to the point of being fraudulent. That's frustrating. When you see, for example, the university so devoted right now to going to class IA [athletics], division IA in football. Everybody has got to attend the football games. [They] make a big deal about it. But I can't get a projector in my class that works consistently. The students sit in broken-down desk chairs. I've got shades in the class that I can't get fixed: they can't be pulled down, they can't be raised; they are torn, they are tattered. You can hardly get chalk. I've got a screen that pulls down and tears. When you have to put up with shit like that, you get pretty frustrated. To that degree you are pretty upset with the powers-that-be, whoever they are.[38]

In [the] physics [department] at Berkeley [where I was a graduate student], no one wants to be department chairman, because it's just routine things. But in departments at universities like this, the chairman controls a lot of things. He controls a lot of money. A lot of research groups get their money from the university through the department. Being in a political position is extremely important in terms of survival, and unfortunately that has resulted in cliques.[39]

One of the three communitarian schools lost its doctoral program in the 1970s. The program had been in place for a short period (only about five years). Many have questioned the need for so many Ph.D. programs in physics (as well as in other fields) when there has long been an oversupply of expert labor.[40] Economies of scale, however, also bear on life courses. For scientists who came to this institution while the doctoral program was in place, and for scientists who predated the program but had since made it an integral part of their lives, the loss of an advanced graduate program brought about decay in parts of their self-identity. Their professional selves adapted to their environments, but not without felt costs.

I would say that some of my aspirations either haven't been met or they have been met only partially. What had a serious effect on me and also a number of my colleagues in this department is the fact that in the really late seventies we lost our Ph.D. program. [A] state board that was governing higher education decided there was too much duplication in the state and they went around and reviewed all the doctoral programs and decided to cut. Not to be in a department where you have doctoral students has been a disappointment for me.[41]

For a small handful of others, this change was of minimal consequence. It is in this communitarian world, remember, that we most often find people who identify strongly as teachers. For those who claimed to have originally embarked on their careers with the primary desire to teach, this world provided a comfortable haven. One sees less change in their self-identities over time than in those who harbored stronger research aspirations. As I explained in chapter 2, teaching careers are relatively unstaged. It therefore comes as little surprise to see only a minimal degree of evolution in self-identity in someone whose career is anchored in the classroom.

My aspirations were to be a faculty member that primarily engaged in teaching with some research on the side. By and large that has happened. That was my hope. I didn't hope to be at a large school, for that matter a Ph.D.-granting school. I would have been happy at a bachelor's-degree school. We aren't a Ph.D.-generating department, and therefore you don't have Ph.D. students. The major difference as I see it is that the Ph.D. student is around for five years and they are involved in the research over that five-year period for four years. You have seminars at the research level. In fact that's the reason that I would say, hey, wherever a physics department can get a Ph.D. program, it is good for that physics department to have that. But I think that there are missions at the master and bachelor degree levels that this university is probably more called upon to do.[42]

Pluralists

Cooling-out proceeded differently among pluralists. In the communitarian world we often heard how scientists made radical transformations in their outlooks. In many instances it was not so much a matter of having to settle for something less: many felt they had to settle for something altogether different from what they had once expected. Pluralists also cool-out, but in the end the process leaves most in a different frame of mind than that of their communitarian counterparts. Most pluralists retain confidence in their ability to proceed with major scientific work. The prospect of achieving greatness, however, has been considerably dimmed. Once their goals were centered on extraordinary accomplishments. Now their plans are scaled to the prospect of doing "solid" work. In Faulkner's terms, they no longer strive for outstanding achievement but have adjusted their sights on average recognition (Faulkner 1974). The people and careers they found on entering this world shaped their outlooks.

I feel that one's abilities and one's skills get sharpened if you are interacting with people who are at a level comparable or higher than yours. I would say that there aren't that many people here whose day-to-day skills are much higher than mine. I probably would have been sharper had I been at a place where there were people who were better than me with whom I had to interact constantly. Here the people are comparable, maybe a little bit better, a little bit worse, but not much higher-level people. When I take a sabbatical and I go to a place where there are people who are very good and better than I am, I certainly find my thought processes getting sharpened, my abilities getting honed. You do see an improvement. From that point of view I would say that I probably did suffer.[43]

I think by the time you're a graduate student you at least have a decent perspective on what kind of structures are out there—what kind of universities work in which fields and what kind of atmospheres they have. There's a gradual narrowing of the pipeline as you get to higher and higher places. You would normally do your postdoc at the very best places. I was at Harvard [for my postdoc but also] had an offer from Princeton. Each place might have two postdocs a year, but they aren't going to have a faculty job in your [specialty] field more than once every three or four years. And so you end up moving a little bit down the stepladder. I think if I were at a place like M.I.T. or Princeton there might be more of a push. I might feel more driven because you're around people who are the best in the world and they let you know it, and it inspires you to work harder. I think this is a more relaxing atmosphere. I'm not quite as driven as I was. As a graduate student, I was very singularly directed towards succeeding in physics, and I devoted a huge fraction of every day to work. I would get up around eleven o'clock, go to work around noon, work until five, come home, have dinner, and then work again until midnight, and then go to bed. I do other things now. I'm married and I have a life outside of my work.[44]

Notice the references to life outside of physics. Familial and extracurricular references are evident in pluralists' and communitarians' accounts, but rarely in elites'. This is especially noteworthy because interview questions were designed to elicit responses exclusively about the career; no questions asked directly about family or leisure. In effect, all three groups were exposed to the same "stimuli" in interviews, but their responses to these stimuli fell into different patterns. Among elites, work is the sole focus of the narrative. All other aspects of life, even those that may impinge on the career, are excluded from commentary.

In Goffman's terms, nonwork sources of satisfaction act as substitutes for a level of identification and reward with work that is not possible in a pluralist world. In abandoning their hopes for scientific greatness, pluralists turn to other aspects of life to compensate for the loss.

My aspirations were to be at a higher-prestige university, initially— Berkeley, Harvard, M.I.T.—types of places like I went to as a student. I was able to get into the best programs as a student. . . . Ideally, I would have, at that time, would have wanted to be a successful scientist at one of those leading institutions. My view of life and career and institutions in general has changed a lot. I think I'm quite content not to be there. It's hard enough at what I would consider a second-tier institution like [this one]. . . . I'm looking for more personal stability now. I want to be able to spend some time on things other than my profession. My family—I just had two kids. That is important to me as well.[45]

Ultimate success is a complete life. It's putting it all together. And physics is just a piece of your life. If physics is your whole life, then you can't have ultimate success. Ultimate success is putting together a job that you enjoy [coming to] when you get up every morning. And the rest of your life: your family, your children, your friends, all the other activities. It's not winning a prize or stuff like that. It's the whole package that comes together. Your wife has a career, and she's doing well, and she's happy in it. Your kids do well and grow up and pursue whatever they are going to do well.[46]

Irony marks the pluralists' reflections. In chapter 2 I described how pluralists constructed a reality that cast their "middle" position in advantageous terms. They viewed their world as a happy medium because they pursued major research without the strained environment they perceived at elite schools. Yet the majority of pluralists completed their graduate studies and postdoctoral training at elite institutions, and thus they had entertained visions of grandeur in the due course of their professional socialization. Many pluralists tell of how they had hoped to be at a school more comparable to those they had known as students. The irony in their accounts stems, therefore, from their adaptations to an environment that falls short of their original expectations. They negatively classify the institutions that they had once harbored hopes of entering.

> I thought I would be at one of the elite research universities. [This school] is not an elite research university. I always tell people that academically, we have a great football team. This wasn't the place I pictured myself in. I initially thought that, if I was the world's greatest astrophysicist, I would end up at a place like Princeton or Harvard or Chicago or Berkeley or some place like that. I think I didn't realize the kind of competition that exists at those places. There weren't any jobs at those places, but that's kind of a Catch-22: probably had I been exceptional enough there would have been a job generated.[47]

> I thought I would wind up at a somewhat better school than this. Maybe something like Northwestern or something like that. They had a research group here, but I had never heard of anybody. It wasn't well-known in my area at that time. It had all the basic ingredients I was looking for. It just didn't have as big a reputation as a place—you always aspire to go to a place with a big reputation. I didn't know much about lifestyle here, except that it seemed maybe not as nice as California, but I had the impression that it had better weather than Chicago, which is true. The traffic wasn't as bad, and it seemed like a nice place to raise a family. The prices were inexpensive. I could come here and buy a house, which I never thought I would be able to do. It started out pretty okay. I was reasonably content in coming here.[48]

As the scientist quoted below illustrates, over time adaptation is in some sense programmatically complete, whether for elites, pluralists, or communitarians. On the one hand, this scientist may be true to his word. On the other, we see Goffman's point: an alternative arrangement has been substituted in ways so cunning as to make the scientist truly believe that this was, after all, what he had wanted. Accordingly, the past appears like a perfect match between the individual and the social environments through which he passed.

> Yes, I think [this is the type of university I thought I would go to]. Yes. Yes, I think so. It was good enough and had the orientation toward research, but it was not one of the top universities. I didn't really feel I had much chance of getting a faculty position at Harvard or some place like that. So this one seemed to pretty much fit what I thought

were my abilities and interests. I think it is true that I have never been rejected for anything up until the time I came here. I had been given offers from every place I ever applied to because I was always pretty careful to apply to places that seemed reasonable for my particular level of talent. Thinking back, it does seem rather remarkable that I always had myself pegged at a level where in fact I was successful.[49]

Elites

Finally, cooling-out among elites constitutes a different pattern still. Entering the elite world, scientists let go of their most majestic visions of grandeur. As earlier passages showed, they saw themselves becoming more "realistic." Hopes of following in the footsteps of Newton, Einstein, and Fermi were set aside.

But unlike pluralists or communitarians, elites continued to hold out hope for making major scientific breakthroughs. The question was not whether they would achieve greatness but at what level. Even though they gave up on fantastic visions (which could, anyway, be excused as youthful idealism), their momentum was unshaken. They were in a world where nearly all examples of careers illustrated the kinds of greatness to be achieved, short of Einstein and Copernicus. They became members of departments considered among the most distinguished in the world. By this overly simple standard, they had risen to the pinnacle of their profession. It is easy to understand, therefore, why many claimed that their aspirations increased in spite of having let go of their most romantic expectations.

> I'm afraid my ambitions have grown. I want to do something good now. You replace one anxiety with another. First of all I had [a] survival anxiety. Now I have the anxiety that I'm given this remarkable opportunity very few people in history have had, that is, to work for some major research university with a secure position. I have no excuse really. If I don't do something really good it's because I wasn't smart enough.[50]

This scientist illustrates a more general pattern. Elites claim that their careers will falter only as a result of limitations posed by their own ability. Their environment is seen as facilitative. It is structurally conducive to carrying out major scientific work. It possesses capable and plentiful colleagues, graduate students, and postdoctoral researchers. Access to major funding and publication outlets is comparatively easy. Modern and technically advanced equipment is on hand, as are technical staff. It is a world in which nearly all of the human, physical, and fiscal resources are in place for major work (see table 2). The burden is thus seen to fall directly on the individual.

The depiction of the world in which great discoveries might be made comes so close to appearing perfect (for those who are dedicated to research) that an alien might regard it as a world of make-believe.

[This school] in no sense has ever constrained my career. There is nothing I have ever wanted to pursue which I could not pursue because I was [here]. [This place] has allowed me to do whatever I wanted to do.[51]

[Being here] has been the biggest plus. A fantastic institution. Terrific colleagues. I've had from the beginning very good support from the department. The students here are terrific. When you don't know something, there is always somebody who can answer it for you. That's fantastic. It's been a great place. I don't want to leave. You know I've been here for a long time.[52]

The communitarian world contained elements of unhappy sentiment, and members voiced their gripes about their situation and age-old institutional politics. Elites tell a different story. The institution is cast as a haven, an unfettered greenhouse for the mind.

I think [the university] has enhanced [my career] enormously because [the school] is a very fine university. They provide me with lots of opportunities to do things; they give me great support. It's been a tremendous plus.[53]

It has not been hard to get resources [here]. The [school's] management is a very convenient one in which to work. I have encountered very little so-called university politics here. I hear tales of some people at other universities, and it's always left me a little bit wondering how I managed to escape it.[54]

Locus of Control

Dissatisfaction with the world in which a scientist works is inversely related to its prestige. As one moves out of the elite and into the pluralist and communitarian worlds, complaints about "the university" increase.[55] A combination of factors most likely explains this pattern, including a scarcity of resources and the governing structures of institutions (Blau 1973).

Moving from elite to communitarian worlds leads to environments of comparatively greater scarcity. Inequities in wealth and scientific infrastructure, as well as prestige, attract and retain different resources, be they human, physical, or monetary. These conditions in turn create different professional opportunities (Crane 1965; Fox 1983; Long 1978; Long and McGinnis 1981; Pelz and Andrews 1966). By the same token, they result in differing states of self-satisfaction. In worlds where resources are highly scarce, we would expect to find political cliques that control the resources and thus the channels for ambition and recognition.

In addition, the governing structures of institutions likely influence the quality of work life. The elite universities in this study appeared to be marked by decentralized systems of institutional control. That is, major authority often was vested in the faculty as opposed to the administrators, such as deans or provosts. This may stem from the authority that individual faculty members command by virtue of their standing in the field. It is difficult for administrators to tell elites what to do. Consequently, decisions are made from the bottom up, and this creates an environment in which individuals believe they control the major forces that impinge on their professional lives (Clark 1968).

By contrast, the governing structures of pluralist and communitarian schools appear to assume more centralized forms. Major authority is vested in department chairs, deans, and higher-level administrators so that control over resources, policy, and personnel falls into a top-down structure. Faculty members are thus in a position to complain not only about a lack of resources but also about decisions, handed down from superiors, over which they appear to have no control.

Intraworld Differences: Points of Professional Divergence

Although careers in each of the worlds fall into three central tendencies, each world is also marked by scientists whose careers deviate from the respective norm. They break away from the typical passage either by accelerating or decelerating in certain phases. These deviations may be thought of in terms of a branching process, whereby an individual breaks away at any one point from the major stem—the typical career path in his or her respective world. The branches of the three major stems sometimes cross and become enmeshed so that the careers of a subset of people in one world resemble the careers of some people in another. This is especially true for pluralists, whose world comprises an array of people whose motivation, commitment, and identification differ: some individuals closely approximate elites; others resemble communitarians; and still others exhibit characteristics unique to pluralists.

Elites

The dominant pattern of the elites' reflections on their pasts consisted of ambition rescaled to more realistic but still lofty targets. This pattern was evident among both the young and the old. Some, however, recall the past as a time of having slowed down or fallen off track. Those who decelerated are normally fully cognizant that their passages represent something different from the norm, and it seems almost obligatory for them to search for clues as to why, if only to achieve coherence in the twists and turns marking their course.

I begin with the account of "Jack Feldman."[56] Feldman is sixty-seven, and he has been on the faculty of the university where I interviewed him for fifteen years. Before that he held an appointment at an elite institution with a less visible physics department. I used parts of his account above, including his early desire to follow a course similar to his Nobel Prize–winning mentors at Cornell: Feynman and Bethe.

His account illuminates the points at which his career diverges from the moral template. Although aspects of his account may be unique, other aspects constitute part of a more general theme applicable to those who wander off, get diverted from, or choose to leave the main path. These aspects include an accounting of the diversionary activities that usually relate to the changed course of career; a slowing down in and sometimes a withdrawal from the research that invariably first led the individual to this world of science; a search for coherence that attempts to explain, or at least account for, perceived or known outcomes of a path followed unexpectedly; and sometimes a search in which the scientist addresses how he or she might recover from a fall or might otherwise compensate for perceived losses.

You said that as a graduate student, you wanted to do important things, to be like Feynman. How have your aspirations unfolded since those graduate school days?

I always wanted to do that. I've always wanted to do, I still want to do, interesting physics that will bring me respect. I'm more aware of [my aspirations] now. I have a whole lifetime to look at. And I'm also very much aware that my life will end in not too long, so that I'm not keen on realizing those aspirations. When I was forty years old . . . before coming [to this institution], I was at the peak of my career. I was going to all the best summer schools and all the best conferences and was giving all the major talks. At that time my aspirations were being realized. I should have been happy. But I wasn't. But I was doing good work.

Why weren't you happy?

Personal problems. The usual thing. In this period of my life I was doing extremely good work. I was doing it mainly by myself. I had some students that I worked with, but all the good stuff came out of myself. I worked long hours at night. The best work I had done was between 10 P.M. and 6 A.M. I remember I would come in at times when I would spend the whole night at my desk absolutely miserable from my personal problems and tremendously excited about my physics and putting out good stuff. Most of my best papers were written under these conditions. I hung on to the physics because the life outside was not so good. So I would make sure that physics kept me sane. That has always been the thing that kept me sane. I just hang on to the physics like a "life ball," because that I know I can do.[57]

Up to this point in his account, work is cast as a refuge. Work was a reality imposing meaning on a life experiencing anomie. Feldman's career at this point was thriving. He was on a rise to prominence, soon recognized by an appointment at one of the world's most distinguished centers for physics research. This period, however, was a turning point in more than one respect. Although he had been accelerating professionally, gaining recognition and fueling his ambition, his career would soon take a different course.

I discovered that I was an extremely emotional person, that I cherished close relationships, that I needed to talk to people, to have friends who I could talk to about emotional stuff, about my relationships and about everything else in my life. Life had been very, very, very dry. Life had been physics and going to parties with always the same people who always said the same thing. It was a big routine where you never got close to anybody. You never talked to anybody about what was bothering you or what could be better or what you loved. That's the way scientists were in those days, most of them still are: a lot of people making very clever jokes, but no one showing a real interest, at least something you could touch. . . . My wife would run away to the end of the earth if I mentioned anything that was different from the usual routine.

What was the outcome of this period?

The outcome of this period was that I got divorced. I found that I was not too late to find out what had been going on in the sixties, and I did.

By which you mean?

I joined all of these groups which I had been wanting to join. This period began to evolve very slowly. Things happened for years that I did not realize, and I became depressed. I got depressed, and when people would ask me, "What's wrong with you?," I would say there is nothing in life for me. And they would say "What?! You were just invited to give talks at this marvelous conference, and you got to stay in this villa in Italy, you can go swimming in the lake when you're not giving your lectures. You can take a vacation anywhere on the face of the earth if you feel like it." That's the way I felt, that there was nothing in life for me.

Did you feel that life was through?

I felt that there was nothing in life for me. Nothing to enjoy. Everything was just another grind. People said, especially my family and friends, that I changed completely, totally. As the years went by, I discovered more and more stuff out there that was of interest to me, which I wanted to get mixed in. I did more and more of that and less and less physics. And there was this period in the middle of the seventies where I did very little physics. I was still invited to give talks, but the talk was running on gas that had been poured in five years before. My physics was sort of coasting. My own private feeling is that I fell off the train. Lots of physics happened in the seventies which I didn't know about. So I stayed behind. When I started doing physics again, say in 1980, one good thing was that my head was completely clear; I could think about new things. Then very slowly I came back. I published some papers on chaos in 1986. . . . I remarried in `91. I'm going to retire in five years or something like that. I think a lot about that. I don't know what my physics will look like. I look at my colleagues who retire; they all keep doing work. They work just as much as they did before. And I imagine that it's going to be the same for me.

Feldman clearly sees an evolution in his professional self-identity, and he explains the causes and consequences of the turning points that structured this unfolding: beginning his work with illusions of grandeur, publishing work that went on to be recognized and that in some mythic sense approximated his idealized visions of self, rising to a prominence that brought international visibility, and then experiencing a set of downward turns at midcareer that were eventually preempted by a resolution of self and by partial recovery.

His life course mirrors a hero's passage in Greek tragedy. The classic form of the Greek tragedy involves rising action or exposition of essential information about the protagonist to dramatize the course on which he or she has embarked. This proceeds to a complication, in which the protagonist is opposed by external and/or internal forces, which then leads to a personal crisis or climax, consisting of a period of high tension and ulti-mately a decisive turning point or set of turning points. In this last period, the protagonist undergoes a reversal; self-identity changes, which may bring about or be accompanied by several other personal and social changes, such as loss of power, social isolation and stigma, and feelings of anomie, degradation, or weakness.

The tragic hero is sometimes said to be undone by a flaw or moral weakness, such as pride, arrogance, or excessive confidence. At other times the hero falls because of a virtue, such as courage, which in the end is thought to be unable to exist in a world of ordinary people. Subse-quently in a resolution of events, often entailing a denouement, the figure comes to discern the complicated events of the past, resulting in a sense of self-growth and newly achieved self-understanding. The hero reaches a self-awareness of his or her life circumstances and his or her relation to a larger order, be it the order of the world or of some smaller segment—work, family—that is part of a life structure. Tragedy dramatizes the idea that people see clearly only when they have been subjected to such great pressures that they are partially destroyed.

Although the details surrounding each scientist who decelerates are in many respects idiosyncratic, they also vary around a common theme: disengagement from work, followed in some cases by a reengagement in which the scientist appears reconstructed, with new voices and visions. These qualities surface in the next account, which was provided by a leading space scientist whom I will call Martin Deutch. At age sixty-four, Deutch recounts an episode that occurred when he was in his fifties. Deutch went on a mythic-like search for his deceased father, presumably to answer questions about or find clues to his personal and family biog-raphy. Taking this path engendered redirected commitments. A question stands as to whether he might have gone further had he stayed in the game of doing science. Yet he slowed down and for a period removed himself from the research front. Progress was made in his field during his absence. He found himself ill-equipped to reenter the field at the level he had known before, and he faced the challenge of how best to apply the sci-entific interests and ambitions that remained. As is typical for a number of scientists in advanced phases of their careers, he moved into textbook writing.[58] Below, Deutch recounts what he was experiencing during these turning points at midcareer.

> I have been tempted to go and do other things. I had a big World War
> Two project that took me off for about two years . . . three years, four
> years. When I was fifty I discovered my dad's letters. He had commit-

ted suicide right after he had gotten back [from the war]. My mother had had a baby by someone else during the war. When I turned fifty, all of a sudden I got really turned on by [all of] that. I guess I was getting to know my father. The letters gave clues to where he was. I started going to the National Archives and getting photographs and journals, and I visited the South Pacific. I went to the Philippines and New Zealand and the Solomons, and even some of the guns and tanks that he wrote about were still there. I put the whole thing together in one book, which has all the letters, plus journals of my writing, my travels. There is something very human about a real story, which has the documentation of a real life. That's history, that's biography. I think people get remembered that way. After I was about eleven, he was off in the army. The reason I got into this [book] was that there was something left behind. You could argue that it was my way of letting grief out. I never broke up, I never became hysterical, I never cried. I was just sort of sad and matter-of-fact about it.
How old were you [when your father died]?
Just turned fifteen, a sophomore in high school. My sister always claimed that I was very quiet. They always worried whether I would kill myself because I was very quiet. Somehow that whole background may have engendered insecurities which drive a lot of us.
Insecurity about what?
How good you are. How well are you doing. Whether you measure up.[59]

Other elites deviate from the typical passage by accelerating in certain phases. They diverge from the moral template insofar as their accomplishments exceed the norms of their age. Their careers may thus be viewed as passing through phases at a faster rate. This also means that they enter subsequent phases sooner than their peers (if their peers reach them at all) and that they are exposed to loftier challenges accompanying each new phase. Professional self-identities evolve in a manner that reflects the expectations attached to newly occupied statuses and the accompanying anticipations "for more." As explained earlier, people not only fill a role but also anticipate playing "the one down the road."

Scientists whose careers accelerated at rates greater than the norm usually made discoveries or had ideas that precipitated recognition in the form of honors, awards, citations, and so on. We see how status and status expectations changed for "Isaac Goldstein," who in the past several years had won a Nobel Prize. Goldstein moved from being a researcher to a spokesman, advocate, and diplomat for science. To some degree his research remained alive, although not on the scale of other Nobelists not part of this sample.[60]

> There are a lot of demands when one wins such a prize. First of all you develop instantaneous wisdom because everybody wants you to be on a committee of some sort. Obviously people know you because you give talks more often, in much greater frequency than in the past. There are a lot of things people want you to do, and unfortunately one has a sense of responsibility, at least I have a sense of responsibility, and it has decreased my ability to work in the past [several] years. I was home for only a few weeks this whole summer because I was invited to a conference to give a talk on an international collaboration in high-energy physics which is something people felt I should talk about. Then I went to Landau [Russia], which is a meeting for Nobel laureates in physics, and I went there because you basically give lectures to students—there were six hundred students there. Then after that I went to a conference in Sicily; I had to go to that one because it was a conference on the history of high-energy physics. So I went to give that talk. And then after that—I'm on a steering committee of the scientists in the ——— Foundation—I went to Colorado. On Wednesday I've got to go to Washington. I'll be there Wednesday until Thursday, and then I'll come home. On Sunday I go to Mexico for a week because I promised to chair a session there. I come home, and then in the middle of October [in about two weeks] I go to New York to do a fund-raiser for a university. Three days later I go to San Francisco to talk to five hundred high school teachers in science. And then I come back from that and on November 5th and 6th, I go off to San Diego for a meeting of the National Research Council Board of Physics and Astronomy. It just keeps on going like that.

Are you happy that you won the prize?

> Yeah, I'm happy. Listen, I'm happy. And this is all very interesting. You just have to learn how to curb it to some extent, which I'm trying to deal with. I've taken the point of view that if anybody wants me to talk to students, I almost never say no. I feel that if I can help in getting young people interested in science, that is exactly what I feel the [Nobel] Prize is good for.[61]

Pluralists

Turning to pluralists, we again see that scientists decelerate and accelerate in various phases. Recall that the dominant pattern characterizing pluralists consists of setting sights on lesser, but nevertheless substantial, achievements. Pluralists dampen ambition in greater degrees than do elites. This pattern, however, does not hold for all. Variety in commitment and identification establishes the pluralist world. In the following accounts, ambition manifests itself in degrees both greater and lesser than the norm.

"Robert Hopkins" has adopted an orientation toward work, and has followed a path, that make his career and his self-identity strongly resemble those of communitarians. Hopkins is now sixty-seven, preparing to retire next year. His early career was marked by a widely recognized theoretic development that resulted in a popular theorem that bears his name. But instead of propelling his career, this achievement began to serve as a career highlight, as it does today. It became his single-most-important piece of work, and his subsequent career edged away from research as a result of deliberate choices he made. He earns respect in his department from colleagues who know him as an excellent teacher with a warm personality, exemplifying the capacity for pluralists not only to tolerate but to admire people with diverse orientations.

In everyone's career there are roads not taken, paths you could have gone down that may have had different consequences for you professionally and personally. What have been the major ones for you?

I think there is one major one. . . . Very early in my life, let me go back. When I first came to [this university], a fellow graduate student of mine was already here on the faculty. We collaborated quite a bit during those first years, and he was driven. He worked and worked, and he set immediate goals and long-term goals, organized very differently from myself. He would push me quite hard. He would expect— not because he was the boss or anything but just because we were collaborating—he would say well why don't we work on this over the weekend so I could get here and here by Monday and then do this by Tuesday, that kind of thing. And I would cooperate to a certain degree, but I certainly didn't burn the midnight oil the way he invariably did. And I made a decision some place in there that I wasn't going to be a twenty-four-hour-a-day physicist, that I have a wife, I have kids, I have a human life to lead, and it's important to me, and I was not so compulsive about achieving the first ranks of physics. That doesn't mean that I've done the best I can within the framework that I've described. I'm a bit lazy anyhow. But that to my mind was a very important decision, because it set the level of my commitment to physics.

From the perspective of a graduate student, is this the type of place where you thought you would go as a faculty member?

As I say, I was rather immature and not very forward-looking. I didn't have, I think, any clear idea of where I would expect to go. I just hoped to land on my feet and fortunately did.

Nevertheless your father, you say, was on the faculty at Columbia, so you must have had some knowledge of how the academic world operates.

I knew that I liked the academic life, that it would suit me well. My father was a fairly forceful, strong-minded person, full of advice, and I was generally fairly docile and would follow his advice most of the time. I don't think it hurt me particularly, but he certainly was a different sort of person than I am.

In what ways?

He was much more of a worrier, a planner. I don't think I ever asked him if he had long-range goals or aspirations, but I have no doubt that from a very early age, he thought very carefully about the directions that he would go. He was very driven. He was very active and more successful, I guess you could say, as an economist than I am as a physicist. He was president of the American Economic Association, things of that sort. I gather he was a pretty good teacher, probably more demanding than I have ever been with students and colleagues.[62]

Similar patterns are evident in the account of the next scientist, who illustrates a casual and permissive, rather than intense and driven, self-orientation. This is a scientist who appears committed to teaching, especially to teaching undergraduates. His research interests slipped many years ago, when he entered the middle phases of his career, but this was apparently not a major concern for him or perhaps for several of his colleagues. He appears comfortable and confident and doggedly assertive.

It has been of no interest to me to be particularly famous. I didn't want to work very hard early and then burn out. I have never had a burning career goal that I had to compete with other people. I'm not a competitive person. I can be just as hard to deal with as anybody on God's earth, and just as aggressive as anybody, but it's not my basic personality. I have no real career aspiration, except to be a faculty member. I enjoy being involved in research; I enjoy being able to teach. I get a great deal of pleasure out of teaching, and it was clear to me that I would enjoy living a life like that, and that's what I've done, and it's been very interesting. I've not been interested in working on Nobel Prizes. I did not set out to be at a certain place at a certain time. I knew generally where I wanted to be in terms of getting tenure. That's the big bugle, right? You want to get tenure and get tenure at the right time. Beyond that I have all sorts of other interests also, and I have always involved myself in those interests in addition to physics.[63]

Like elites, pluralists also deviate from their central path by accelerating at various points in their careers. Often such pluralists exhibit several characteristics typical of elites. They have been heavily active in research and are often recognized like elites (see table 6). "Philip Stockman" is such a person. He spent several years working in industry before accepting an academic position. Stockman's ambition is most clearly expressed through the research that he does. He has been a highly active member of the physics community throughout his career, and his account testifies to this commitment.

I really enjoy doing research. That's why I work. I'm sure you'll find that most people who do research don't work forty-hour weeks. I work sixty-, seventy-hour weeks because I like what I do. I go in the lab and get the experiments done. [Having] a lot of drive [is important]—a sense that you want to accomplish something and a willingness to constantly work hard at it. More important than creativity is the consistency of effort—[that's] what in the end makes the difference. I've seen too many people who are very creative who do an occasional very good piece of good work, be it in applied research or in technology or in fundamental physics, but then don't carry it through; they don't promote what they are doing to other people, they don't talk to people, don't build on it. And when there is a setback, [they] take it too personally. There are always setbacks. I've been fortunate enough to help start a field of plastic metals and founded the field of plastic magnets. Both are doing well, and I'm pleased to see them flower, to try to make them go forward. I take responsibility for organizing conferences; I enjoy that. I just organized a major one for the field of plastic magnets in Salt Lake City. The new science continues to be exciting, and it is also exciting to see the transition to technology. In the end, society supports the science we do at a higher rate than they support the arts because they [society] expect a spin-off from the science. I don't find that bad. The people who pay the bills deserve to get something for it, and I would like to see the science we have developed, both what we develop in our own lab and the general field I'm in, pay off for society. And that statement extends to the fact that I also think it's my job to make sure that the undergraduates get trained well too. I have some loud discussions with my colleagues sometimes who feel it's beneath them to take seriously the undergraduates who are not very interested in the science they teach. They are your cus-

tomer. They are paying part of the bill, and you [as a researcher] benefit from them paying the bills. They deserve your attention. I have
achieved many things I was hoping to achieve at this point. I think the
goal that I mentioned about seeing the science flower and seeing the
commercialization of it, the implementation of it, I'd like to see it occur, and I work toward those ends.[64]

Communitarians

In turning to communitarians, we see scientists who similarly deviate
from the customary path, though this path emphasizes modest scientific
achievement. In various career phases, some communitarians accelerate
or decelerate. Where they decelerate, research has normally been aban-
doned altogether, for this is the world that places the least emphasis on research: the master self-identity consists in being a teacher, almost
exclusively of undergraduates.

Those communitarians who accelerate normally have lengthier publication lists, and they are often more active in attending meetings. (Such
communitarians also have higher salaries than non-research-active
peers, as is true also among pluralists and elites.) On the whole, their
scholarly records are similar to those of the many pluralists active in research. Rarely, however, do the research records of communitarians bear
strong resemblance to those of elites, in large part because the cultural
and structural conditions of productivity differ in considerably greater
magnitudes between these worlds (see table 2). The careers of research-
productive communitarians are best viewed as miniature versions of
elite careers.

Aside from conducting research, scientists also make significant status
gains—whether as communitarians, pluralists, or elites—by embracing
administration. Except for a brief mention at the conclusion of chapter 2,
I have not yet discussed administration, mostly because few scientists
with much in the way of any administrative career were included in the
sample.[65] Administration is not an institutionalized stage of academic
careers, as it is for, say, engineers (Zussman 1985), who may be considered to have failed if an administrative role is not assumed by midcareer.
Thus, it is not unusual to find so few administrators in this sample: embarking on a study of science careers, one expects to find a majority of the
scientists invested (albeit in varying degrees) in activities that define the
scientific role—research and teaching.

"Robert Fairweather" is one scientist who has devoted several years to
successive administrative positions. While rising in increasingly power-

ful administrative positions, Fairweather worked his way down through institutions of lower and lower prestige. Administrative ambition possessed him, and he apparently found his way to the places that would accommodate him and his ambition. He started off in a pluralist world as an assistant professor, first at the university where he received his doctoral degree and then at a nearby university. He then spent several years both as an industrial scientist and as a visiting professor.

Ten years after receiving his Ph.D., he became associate and subsequently full professor at a pluralist institution much like, for example, the University of Kansas or Missouri. After five years he moved to a similar institution as professor and chairman of the physics department. After another five years he became provost and vice-president for academic affairs at the school where I interviewed him, never having been a dean or having served in some intermediate administrative capacity. When I spoke with Fairweather, he had "been out of administration" for five years and had assumed a full-time professorial role in the department of physics.

In taking these administrative posts, Fairweather became highly active on the national scene in science. He served on panels of the National Science Foundation in Washington, D.C., and made regular trips there from his various academic homes. He became involved in state education issues, promoting science education on state boards and in the legislature. He sat on a number of review and accreditation committees, and he organized and chaired sessions at physics conferences. Like many other scientists, he was asked to be included in *Who's Who in America* and *American Men and Women of Science*.

While his administrative responsibilities grew, he kept a hand in research, although his productivity was tempered by the burdens imposed by these administrative duties. Apparently he maintained a record of research throughout this period in large part because of the collaborative nature of his specialty and, indeed, of much of physics and other sciences. Where he was not the point man, he was part of a pack.

Fairweather is highly personable, as one might expect of administrators and also of communitarians. He is relaxed and confident—happy, it seems, to have relinquished his duties as provost.

> I think ego got the better of me. People said, hey, you have great leadership ability, you can do this, that, and the other thing, and the next thing I knew I was the provost of this university, and I had not a lot of administrative experience. The problem for me was that I had to spend all my time shaking people's hands and making people feel

good. I spent less and less time actually accomplishing something. Why they hired me I don't know. At the time it seemed like a real ego trip.

What do you think motivated you to come here as provost?

I think it was just ego. This was an opportunity to have control over a $60-million-a-year budget, make an impact on the university, and I think I did make an impact on this university. There are a lot of things that are different that I started here.

From the perspective of a graduate student, is [this] the type of university where you thought you would end up?

I've wondered about that a couple of times in my life. A lot of people are much more conscious of that than I am. I've moved a number of times, always where there was some advantage to me in doing what I wanted to do. So, for example, there is a big advantage in [coming to this state] over any other if you're at a public university, because [this state] has a lot of money. I got a lot of money when I came here to work on research, and I still get a lot of money from the university for work on research, more than I would have gotten [at the two previous institutions where I worked]. My salary is also quite a bit higher [between $80,000 and $89,999]. I make a very good salary. I have a son who graduated from Berkeley as an undergraduate, and he's now at the University of Texas at Austin, and he wants to be a professor. He is extremely conscious of where he's going. Austin has one of the top-ten programs in computer science, and that's why he's there. He wants to be at a prestigious university. That's very important to him. It's never been an issue for me. My issue has always been what can I do, what is open for me at a particular place. I didn't go to USC [University of Southern California] because, while it was a very prestigious kind of place, I guess, they didn't have any money. We have a lot more money than they do. I didn't go to Santa Cruz [University of California–Santa Cruz] because that was like a whole [program-]building experience. Maybe I'm short-sighted, I don't know. I feel pretty happy; I've been pretty happy most of my life.[66]

Conclusion

I have described how the past appears to scientists belonging to distinct academic worlds. In the course of doing so, I have attempted to draw a picture of the outlooks formed by scientists after making a series of career adaptations, in particular those adaptations that occurred as a result of the unfulfillable expectations they brought to their careers. Their accounts tell a story of moving from youthful idealism to what they believed to be more and more realistic appraisals of self. But these appraisals were rendered within different environments, offering different opportunities for a scientific life.

The unfolding of careers brought about changes in professional self-identities. Scientists began seeing themselves, and their prospects for achievement, in new ways. They gave their visions of the future a second look and, nearly always, found they had to make major revisions. The transition from school to work and on into the career involved rewriting scripts that they would attempt to follow. The rewritten scripts were styled around the opportunities they found, and the expectations they faced, in their worlds.

I have postulated three pathways that characterize typical passages in each of the worlds. These may be seen as distinct "cooling racks" on which people adjust in the heat of trying to do science. Elites hold out ambition for greatness. Pluralists settle on more modest, but nevertheless substantial, achievements. Communitarians hold still more modest achievement expectations. Professional self-identities evolved in ways that reflected the opportunities available in the worlds in which the scientists worked. If the ambitions of the scientists appear to vary between the worlds, it is because these worlds present variant opportunities *for* ambition.

In this regard, a place becomes a symbol of one's talent and mobility prospects. It operates as a metaphor for how far one can travel. In the minds of scientists (as well as in the public mind), different places impose different ceilings on the reaches of potential. For the majority of scientists in each of these settings, place was a self-fulfilling prophecy. They began adapting by finding a locally feasible script and by speaking its narrative. But for others, place acted more as a myth.

Whether complying with a pattern or deviating from it, these career experiences strongly suggest that high-profile success prompts more anxiety and relentless drive than do low-profile performance and the de-escalation of dreams. Success works like a drug for a disease that proves to have no cure: the more success one has, the more one wants; the more ambitions are satisfied, the more they grow. It may therefore seem ironic to assert that the most satisfying careers in science are those in which downward mobility began at the outset, leaving scientists satisfied (in the end) with alternative career paths because they were never lured by illusions of grandeur once past graduate school. By contrast, for those whose ambitions mount, there is no rest—and ultimately no passage toward completely satisfying their ambitions. They internalize a community mandate that operates as a kind of reverse Matthew effect: from those who have given, more is expected.

For communitarians, the prospect of contentment is real and local but always diminished by the wider, cosmopolitan career lived out by elites. For elites, worldwide recognition is real enough, but the prospect of contentment is diminished by local comparisons with others of equal or greater recognition and, if not with them, with the pantheon. Even when all other forms of recognition affirm an individual's grounds for satisfaction, there are always reminders of one's incompleteness: communitarians are reminded by salaries that are adjusted to favor research over teaching; elites are reminded by awards, prizes, endowed chairs, and so on that often bear the names of people greater than themselves; pluralists, wedged in between, are reminded by combinations of the same things.

We have come to the point in this study where the scientists are living out their careers. Our sense of them as situated in three distinct, though partly overlapping, worlds has been cultivated. We have heard of the adaptations they needed to make in these worlds. We have seen what opportunity looks like on the inside. To complete the picture, we must join visions of the past with those of the future.

FOUR

Professional Futures

In the previous chapter I discussed the ways in which scientists accounted for their professional pasts. Yet the process by which individuals create professional self-identities is dynamic. Self-identities are not fixed in time but are routinely revised through a process in which individuals, at any given point, account for the past in light of a desired future. To examine only how these scientists account for their pasts would therefore result in an incomplete view of how they form self-identities. To gain a fuller perspective, we need to hear not only where they have been but also where they see themselves headed.

This chapter will present the latter accounts. I will be primarily concerned with how the scientists conceive of a *future self*. The future self is an image that individuals hope (and perhaps sometimes firmly believe) they will embody at a later point in time. It is an imagined identity, a constellation of ideas of what one fantasizes about becoming. We saw examples of future selves in the scientists' accounts of where they saw themselves headed when they were first starting out in science. Many had modeled images of themselves around mentors and physics icons.

The future self has special salience for physicists, but by no means is it exclusive of them (for example, see Schmitt and Leonard 1986; and Levinson 1978, 1996). Physicists (and other academics) hope that their work is long-lived, achieving a kind of immortality. There is a creative element to their work, much as there is in other arts and sciences. But for scientists, the quest for immortality stems from the way in which the science community sanctions progress, namely through recognizing individuals.

The occupational practice of recognizing individuals is especially remarkable in light of the communal system in science, in which joint efforts generate knowledge. Major (and even minor) achievements in science are often connected to the name of the person or persons whose efforts *directly* resulted in the achievement: Einstein, Newton, Copernicus. Although these are among the names that are best known outside of the science community, much smaller achievements also often result in immortalizing individuals: the Yang-Mills theorem;[1] the Chandrasekhar limit;[2] the Josephson effect;[3] the Richardson number;[4] the Thomson coefficient;[5] the Geiger counter;[6] Faraday rotation;[7] the Debye-Hückel theory;[8] the Wheatstone bridge;[9] the Weber-Fechner law.[10] Such people (and their technical accomplishments) may be recognizable only to fellow scientists. But the practice of assigning names to achievements highlights the pattern of establishing an eminent identity, a pattern that most physicists hope to emulate. Following the dictum of the ancient Greeks, fame is the closest one comes to immortality. Scientists' concern with their future selves is intimately tied to the place they aspire to achieve in history.

My focus on the future self is motivated by a theoretic objective that guides the overall study: the role that ambition plays in a career. By definition, the future self is an image around which an individual orients his or her actions. It is a source of motivation that draws its energy from anticipation. The effects of the future self on the emotional state can be diverse, both among different people and within a single individual. On the one hand, the future self may be inspirational by directing efforts toward noble goals. It may also function as a release from tension by providing a temporary escape into an imaginary world in which people see themselves basking in the afterglow of pinnacle performances. On the other hand, the future self may engender feelings of inadequacy by unsparingly pointing out the disparities that lie between the imagined and the "real" selves (Turner 1976). In this manner, the future self may be discouraging for people who at times feel that their efforts, however strong, will never result in realizing their most cherished hopes.

Future selves help us understand the forces that drive people in their professions and thus represent a main path to discovering what makes scientists tick. In keeping with our overriding questions, I want to know how scientists are driven and in what magnitudes and intensities. I base my discussion on a set of five questions I asked the scientists:

What do you dream about in terms of your career?
What ultimate thing would you like to achieve?
How do you envision yourself at the end of your career?
How would you like to be remembered by your colleagues?
What about your life do you think will outlive you?

A useful way to ground the discussion is by interpreting the scientists' accounts in terms of the moral career first described in chapter 2. Given the circumstances in which scientists presently find themselves, they are headed in directions with options—some limited, others broad. The intent now is to see in what ways their prospective outlooks remain consistent with, and in what ways they diverge from, the moral career.

At times in the last chapter the scientists' narratives may have seemed prospective. If at times in this chapter the narratives seem retrospective, this is precisely due to the manner in which all people construct the present: through a mediation of past experience and future expectation (Mead 1932). Expectations for the future play out in the present by taking account of the past. In like ways, conceptions of the past are shaped by articulations of the future. It thus would be difficult to gain the proper perspective on any dimension of time without stepping over the cognitive boundaries that help us order our passage through space. Though this and the preceding chapter are distinct analytically, the self arises (and is understood) only through temporal interchange.

I will first turn to differences between each of the three academic settings. I will begin by discussing the future outlooks of members of the two worlds with greatest contrasts: elites and communitarians. A discussion of pluralists will round out the picture. I will then turn to how perspectives on the future differ among scientists within each world. As in chapter 3, my strategy is first to provide an account of dominant patterns and then to examine the ways that individuals sometimes break away from their respective central paths.

Interworld Differences: Being and Becoming

Elites

My earlier account of the moral order among elites captured the sense of activity that fills the atmosphere, defining their character. The pace of occupational life appears swift and vigorous. In my interviews with elites, their answers to questions about their futures resonated with special meaning. Much of their self-identities is staked in what lies ahead. Elites are most often "future-oriented." What is notable is that this concern with "what is next" figures as much in the accounts of the young as in those of the relatively old. Careers are constructed with language that calls attention to what is waiting: the next project, the next article, the next experiment, the next talk, the next proposal.

This is a world in which the self is in a perpetual mode of becoming. Self-identities change as accomplishments are reached throughout the course of a career. The rates at which elite scientists advance differ over the course of their careers, which I will discuss fully later. Patterns vary:

some scientists may gradually slow down while still making progress toward desired ends (a leveling off on a linear upward slope); others may carry on in fits and spurts (an upward-tending jagged line). Regardless of the pattern, the self in this world is an object that rarely comes to a point of rest. Those who do rest are thought to be exceptions to the more general rule. The call to achievement in this world does not cease with the comfort and security supplied by such conventional milestones as tenure and promotion.

We see the image of a future self in the account of "Geoff Silverman." Silverman, who is at midcareer, eloquently describes an identity he hopes to assume by demonstrating scientific prowess.

> The dream is to discover some fantastic new effect that knocks the socks off my friends and colleagues, that knocks the socks off the community, so that when I walk down the corridor, the young students know me and say, "There goes [Silverman], he invented the [Silverman] effect." That's what I want; I want my effect. I want to be the first person to predict such and such an event and for it to be . . . I can even smell what it's like already. It has to be something which once you think about it, is very reasonable. Very surprising at first sight, but at second sight, yes, of course, that's how it had to be. I want one of those. I want my Josephson effect, my fractional quantum pull effect.[11]

Future selves can hold multiple functions for individuals. I asked Silverman what functions his dream held for him.

> It's probably slightly negative most of the time. It's such an unrealistic goal that I probably, if I ever get discouraged, which I don't very often, but if I ever get discouraged, having that dream adds to the discouragement, because the time available to achieve that diminishes as we speak. It's probably something like wanting to win Wimbledon. The older you get, the less likely you are to do that.[12]

But notice how he holds out hope.

> But I've got a few more years. I think the best tennis players are the ones who focus on the next match, not on winning Wimbledon, and probably the best physicists are the people who focus on the next equation and not on climbing the steps at Stockholm. We all dream about walking into the travel agency and asking for our tickets to Stockholm. We even joke about it. We talk about those colleagues of ours who get twitchy every time the phone rings in October, because

you know that's when the phone call comes—something we openly discuss, most of us. I probably think that anybody who says they don't want the Nobel Prize is probably lying. We all want that ultimate seal of approval. I certainly do. It will give you the opportunity to thumb your nose at the few people that you've got tangled up with over the years.[13]

Becoming—assuming successively greater statuses—is not limited to the actions that people take, or the courses that they follow, while living. Death confers a still different status. Death functions not only as a capstone on a life but also as a springboard to an identity made glorious by acts and pronouncements of valor that sometimes come only in times of death. Death, therefore, does not end the process of becoming but is an additional stage. Death involves rituals that call forth public (and private) statements that clarify and install in the collective memory how an individual is to be remembered. Remembrance may be short-lived (except, of course, by friends, kin, and select others), but the transitional process from life to death involves establishing a legacy that the individual may have worked hard to produce, however lasting (or not) it may prove to be. Memorial services and obituaries dramatize the life and times of the dead, often highlighting the individual's unique abilities, accomplishments, and qualities. Many scientists will never see fame in their own lifetime, but in death or long thereafter, their work and their scientific identity may take on greater significance.

Below, Silverman offers an image of how he wants to see himself, and how he wants others to see him, at the end of his career. Interestingly, he equates the end of his career with the end of his life by invoking post mortem references. His account reveals the role that quests for discovery play in how elite physicists form outlooks on themselves in the future, even in a future in which they have no living part. Moreover, the account emphasizes the way in which future selves speak of a place that one hopes ultimately to take in history, emphasizing this dominant theme. Notice the adjectives that Silverman uses to describe himself and his work, which he hopes will stand the test of time: good, unusual, fantastic, adored, brilliant, beautiful, elegant.

I would like to be remembered as a good teacher, which I think is quite unusual, but I really care about classroom lecturing and I love doing it. I'm good at it. I read the obituaries in *The New York Times;* it's the first thing I read every day. I want one, and I want it to say that he was a fantastic teacher; his students adored him, and he taught a generation of physicists. That's one thing, and the other thing I want is: he was a brilliant researcher. He invented Effect X. I don't want the 500 publications, like the people in the biochemical communities have. I don't want those things. I'm happy to publish three or four, five or six papers a year. I don't want big grants. I don't want money. I want to do something which is a subchapter in the next history of solid-state physics. It doesn't have to be the whole chapter. It's not like discovering superconductivity. But I want my own section "5-point-something." And it has to be beautiful. I don't want to find something that's just hidden away and if you dig deep enough it's there. It has to be elegant, and it has to be describable using very elegant pictures of mathematics.[14]

Although the specific aspects of future selves differ from person to person, they vary on a common theme: the theme of self-striving. In the next two scientists's accounts, we again see the forms that future selves take in relatively young physicists. In the first example, the physicist, age thirty-two, has just entered his second year as a university professor. In the second example, the physicist is thirty-three and in his third year of teaching. Both scientists describe how they envision themselves at the end of their careers.

My fantasy [is] that I would be in my sixties or seventies and I would have such a luminous set of interesting things that happened in my scientific career that I would give after-dinner talks and I would go on public lecture tours and would be out saying that the world is fascinating, it does this and it does that.[15]

I would probably be recognized. Right now I am already recognized by the community, and then [at the end of my career, the question will be] how big a leader. That's how I would envision myself. I hope I will be recognized as a leader of one branch of theory.[16]

For both scientists, future selves frame the ways in which they hope to be remembered, with a presumption (or even outright declaration) that great memories come from securing a place in history. The passage below is taken from the account of the second scientist, who talks about what he hopes will survive him.

A foundation of a theory—anything like that would outlive myself. But then the question is magnitude, a bigger theory or small theory. The people at [this university] all have a significant contribution, and some of their work will outlive themselves.[17]

This passage also testifies to the dominant pattern, discussed in the previous chapter, among elites: they continue to hold out hope for greatness, even though their ambitions have normally turned away from following the founding fathers or time-honored heroes.

In moving from the young to the old, the key difference is the object of ambition, not the existence of ambition. By late-career phases, normally when physicists are around their mid-fifties, we see changes in how future selves are set. An emphasis on the process of becoming remains, to the extent that this process involves continued professional development and activity. But the future self contains less of the idea that one will become something substantially different. Older physicists have come to the realization that they have gone just about as far as they will go. They believe any status change will, in all likelihood, be small: their careers have peaked.

The images they offer stress continuity. Older elites do not aim for greatness. For several, greatness has already been achieved; now it must be preserved. Even those older physicists whose claims on greatness may be meager or nil still mention "being busy." Like their younger counterparts, older elites answered my questions about career fantasy. In the following example, we hear from a fifty-nine year-old physicist who comments about his dreams.

> Basically, to be provocative scientifically. I get many invitations to meetings, to give invited papers. That is distracting. It's also rewarding in a sense— that the international community recognizes your work. A little bit of this recognition I would like in the future. Not to get as many invited papers as I get now, but a few, five years from now. That would satisfy me.[18]

In his comments on "ultimate achievement," we also see striving. Grand goals make a life seem grand.

> To build experiments for the proof of the feasibility of fusion. But if your student does it instead of you, it is not separate. I would like those goals to be achieved, but I would be glad if they were [realized] by my students or collaborators.[19]

Similar patterns are found in how he envisions the end of his career. Activity forms the root image. His recent and upcoming schedules (nearly everyone I talked with seemed to have a "calendar") shed additional light on how this man imagines his career on the horizon.

> I will continue writing papers and listening to talks. The things that I would fear most is the ability not to travel. Now I travel too much. But there is so much enrichment in talking to other people. I'll give you an

example: since the end of the year, May 15th [it is September 14th], I
have been in Europe four times and one time all the way around the
world. I went from [here] to Paris and then to Montpelier. And then I
went to Tokyo. Next week I have to go to Florida to give a talk on astrophysics. Then I have to go to Seville, and then to Rome, and then I
will be back—all within ten days. So at the end of my career, I would
like to keep doing that. It would be hard not to be able to go [places]
where there is science being done.[20]

Indeed this depiction of his career's end appears so alive that one might
question whether this physicist envisions an end at all.

At the age of sixty-six, another physicist explained how his perspective
on the future is anchored in the discovery of major concentrations of hydrogen gas in the early universe. He claims such a discovery would lead
to an understanding about the early universe that no prior histories of the
galaxies have accomplished.

I know what to do if we detect a [radio] signal in some direction; we'll
want to make a more intensive study of it. We'd like to find more than
one example of a massive concentration of pregalactic hydrogen.
Many others would probably come thundering in. That's the sincerest
form of flattery.[21]

Elites were frequently ambivalent about retirement. They often in-
sisted they would never retire, as if retirement would cut an aortic channel nourishing life. Even those who said they might retire envisioned
schedules that differed little from those of the present. The senior woman
physicist quoted below is typical.

How do you envision yourself at the very end of your career?
    Maybe I'm there now.
You feel you are there now?
    Well, I'm sixty-three, almost sixty-four, so I'm getting to that point. I
    plan to keep working as long as [my] health is good. This is fun for me.
    I like it.
Do you ever plan to retire?
    I'm saying after I officially retire I just plan to keep working.
Do you have a sense of when that might happen?
    In the next few years. I don't think my lifestyle is going to change that
    much.
Do you know what kind of routine you would like to have?
    Same routine. I know it sounds ridiculous, but I enjoy it. Nobody
    tells me to work as many hours a day as I do. [I'm] usually here at

six o'clock or a little after six o'clock in the morning and then I leave at about 7 P.M. Then I usually do a few things in the evening after dinner. It was just like I was as a graduate student. I haven't changed that much. I work many, many hours during the week. I do it because I enjoy it.[22]

Both continuity in behavior and ambivalence about retirement are seen in the account of the next scientist. The future self of this older elite is fundamentally the same self he sees today. He has likely peaked in his career, and change in his status will probably be small. Nevertheless, his future appears vested in activity.

What do you dream about now in terms of your career?
   I think just [to] continue what I am doing—make successful proposals for guest observations [of the solar system]. And write papers about the results.
Is there something you would like to attain, an accomplishment you would like to make?
   No, I think just do as I am doing [now]. I'm now sixty-six. I imagine that I will formally retire at seventy, provided I can bargain for an office and a parking permit, because I intend to continue doing what I'm doing. The retirement provisions are very generous indeed. I could end up with a higher income when I retire than I presently have [which is between $90,000 and $99,999]. It's not an economic problem. I think it's reasonably important to retire at seventy to free up the salary for younger people.
How do you envision yourself at the end of your career, whenever it may be?
   Well, hobbling in on a cane and sitting down at my computer. Still being active. I have no desire to retire to Florida.[23]

Many scientists see added, rather than fewer, attractions in retirement (or quasi-retirement). The physicist above spoke of the financial rewards. In addition, retirement can grant greater autonomy. This situation creates ironic conditions: scientists engage in efforts with even more vigor (and less distraction) than before.

   I probably wouldn't have as much teaching obligation. After retirement, if you still have your marbles and are still effective teaching, you can teach up to half-time and be compensated for that, in addition to your retirement [income]. I probably wouldn't want to teach as much; I do enjoy my research. I had a sabbatical leave last spring and this summer, and it was just great to be able to just focus on that exclusively and not have committees or students to worry about.[24]

Communitarians

Whereas the dominant trend among elites consisted of *becoming*, the leading trend among communitarians consists of *being*. Communitarians' images of their future selves capture a measure of growth in early phases and then level off. In the early phases of their careers, young scientists have yet to hold secure positions, and their more romantic dreams are still being cooled. The future selves in these phases differ the most from how people see themselves in the present. Following these phases, individuals often appear to lose momentum. They do not keep pace with their elite or pluralist counterparts (see table 3). Communitarians are quicker than elites or pluralists to realize that they have gone as far as they will ever go.

These differences may be understood by recognizing how the life course is socially controlled. The life courses of elites come under comparatively tight and stringent social control. Individuals are pressured to perform at high levels through the duration of their careers. Those who deviate typically must confront, and ultimately come to terms with, various negative sanctions. By contrast, among communitarians, social control in middle and late life is less evident. The most pressing social control applies to the master role of teaching: failure to satisfy assigned duties brings sharp reprimand, but "failure" in other domains is tolerable.[25] A communitarian has to be conscientious but not compulsive.

Compare the future perspectives of the young elites quoted above with those of two young communitarians. Both of the latter talk about how they would like to be remembered.

> In trying to answer your question, I'm thinking not about myself but about older people who have retired and how I respect them, how I think of them. I just want to be remembered as someone who does things, who is not a crook, someone who just lies around and does garbage types of things, monkey business.[26]

> I don't know. I don't have any super-special desires or anything. I would like to be remembered, you know, as a guy who worked hard and, you know, did a reasonable job, and that's about it.[27]

These perspectives are forward-looking to the extent that they refer to personal characteristics that can move one ahead (being active, working hard). But these visions differ from those in the earlier accounts: young communitarians set their sights on closer horizons. For the sake of argument, we can assume that they are already active and hard-working. Yet if these scientists died tomorrow, their lives as envisioned would already be complete. They impute to their future selves characteristics they already possess.

Even more striking differences between elites and communitarians are found in middle and older cohorts. As communitarians age, their interest in science wanes significantly. Evolution in identification with and commitment to science cannot be explained merely by a passing loss of interest that strikes scientists of all stripes. The scientist quoted below is sixty years old. I asked him how he envisions the end of his career.

> Studying physics hasn't really moved me much further here from where I started, meaning that what drew me into physics at first, I thought well, there are a lot of interesting secrets in nature that could be discovered and that this would give me a much better understanding of the world around me and all that kind of stuff. But I find now that physics has not provided those answers or that satisfaction, and I don't think it ever will. One has to question, what's really the point of it? I might decide that I don't want to be active in the day-to-day research program, rushing down to the lab, putting this together and that together, and so on. Cranking out results. I guess I'll probably give up doing that. I'm not sure exactly what I will do really, to tell you the truth. I might get interested in peripheral things. I find computers kind of fascinating; that's something that I didn't have any exposure to while I was a student; it's all new; it's like a toy. That interests me. And I may find something else that interests me, but it probably won't be what I've been doing for the last twenty-five years or so. It will probably be something different.[28]

Leaving science altogether after a long career is, of course, viewed by the science community as a legitimate exit. For the majority, a "cold" exit is the norm.[29] We have heard the several ways in which physicists view the end of their careers: some careers never end, some end slowly, and some end abruptly.[30]

A phenomenon found primarily among communitarians is that of *getting stuck*. For many, science fails to take them where they had once hoped to go. Getting stuck brings about ambivalence. A dialogue with a sixty-two-year-old scientist illustrates this pattern.

How have your aspirations unfolded over time?
> I was [department] chair for five years, and I knew then that I didn't want to have anything more to do with administration. I think that I went through a fallow period of research when I didn't do much in the way of research.

What period was this?

That was when I was chair and then afterwards. I have gotten back to doing, not as much as I probably should, but I'm keeping reasonably active.

How did this period affect your work?

As far as the day-to-day operation of the department, or even my teaching—and I still had a student—that really didn't suffer that much. But I wasn't, I just had lost interest and I wasn't as aggressive.

What do you dream about now in terms of your career?

That's a problem. That's a problem. I don't know. I just don't know what I'm going to do. I don't know.

In what ways is it a problem?

Well, I mean, I can see ultimately stepping down [from teaching], so I think my self-identification as a physicist is rather complete. I read, I play tennis, I bike, I do other kinds of exercise things.

How do you envision yourself at the end of your career?

I'd like to get another idea. You have maybe one or two ideas in your life, if you're lucky.

How would you describe the period that you are in now?

Approaching fallowness again. I don't like it; I don't have any ideas. Maybe what I probably ought to do is either take a leave of absence or retire or a very long sabbatical—maybe a combination of a leave of absence and a sabbatical and go somewhere else and try to learn something else. I don't know if I'm smart enough to work on this [experimental work] because I don't think I have the energy to do experimental work anymore. That requires an enormous amount of energy and enthusiasm—to get an experiment to work.[31]

We see similar patterns in the account of a scientist who is only fifty-one.

In about four years I can retire. . . . I can sort of see myself staying here for about four more years. Theoretically I could retire now. In the retirement system, you can quit after twenty years. You can immediately get retirement salary. But there is something magical and mystical once you pass [age] fifty-five because the way they compute your retirement—[it] gives you like an extra 12.5 percent. So I could see myself lasting for another four years. But then if I do that, or even if I were to leave before then, occasionally there are thoughts that have crossed my mind. For example, if I had enough capital I could see myself going elsewhere and opening up a bookstore.[32]

Entrepreneurial interests often accompany getting stuck. Such interests sometimes involve ventures that would remove scientists from their roles as professors, as the scientist above makes clear. In many instances a scientist's technical training is not applicable to the new venture (for example, opening up a bookstore, playing the stock market, or joining a real estate brokerage). In other instances, the technical aspects of science—not exclusively of physics—are applied in technology development, consulting, and other envisioned roles. Sometimes individuals believe they will enter these roles while simultaneously keeping their university ties, and at other times the pattern parallels the one cited above: they contemplate leaving academic science altogether.

Entrepreneurial themes are confined to communitarians' narratives. A scientist more at home in the elite or pluralist world might scarcely know what to make of these extraprofessional references. Because these findings are not isolated to one or even a few individuals, they cannot be dismissed as exceptional. They represent a pattern, which says more about the social world in which these individuals work than about the individuals themselves. What it says, I argue, is that this world provides comparatively fewer opportunities for ambition. The consequences of limited opportunity are seen in how scientists adapt their careers to their environments: they decelerate and often lose interest in their work altogether.

The next scientist's outlook on the future is cast as one of multiple and highly variegated possibilities. He is in the early career phases and has remained active. Nevertheless, his commitment and identification do not lie solely with science.

> I've started considering small businesses and ways of making money by devising products, and I think that might be an interesting alternative . . . something like writing software or applying some of the stuff in the lab—applying diamond film research to making new products and selling them. Also just doing something totally different that's more in the mainstream of the economy, like selling insurance or the stock market, something like that. I haven't really been getting more involved either in the stock market or anything else, but I have thought about it as an alternative, mainly because the monetary rewards are so much greater. It would have to be a full-time job because you have to know what's going on, you have to read the newspapers, you have to see how the stocks are doing. It's a full-time job. That's what everyone tells me. I may very well do that because you can make so much money. Probably a six-figure salary would not be out of the question if you devote eight, ten hours a day to it.[33]

Another scientist, in mid-career (age fifty-two), formed a similar outlook, one reactionary to the communitarian world.

What do you dream about in terms of your career?

Forming a company, taking it public, making a number of millions of dollars. I don't know. There is not much more I can do here, I don't think. I am involved in companies now.

In a consulting capacity?

I do that as well, but as a stockholder. I have been involved in companies before. This was a private company that was formed to, back when the oil business was going well, back in the late seventies . . . this company was designing software to go with the down-hole tools. These are tools that are put down an oil well and radiate the formation, and then [we] look at what comes back from the radiation to tell whether there is oil there or not. First I was going to be given one-third of this company, and we would give 80 percent of our stock to [another company], and they would give us a comparable amount of their stock. I was on a retainer; I was paid quite a bit of money, plus given a car to drive because I had to drive to [an adjacent city] a couple of times a week. That was just a lot of fun. It was a brand-new deal.[34]

Pluralists

More so than elites and communitarians, pluralists follow diverse paths. Their outlooks sometimes resemble elites' outlooks, sometimes communitarians'. Once again, though, we see that pluralists generally bear more similarity to elites than to communitarians.

The central trend among pluralists is a striving for strong, solid (but not exceptional or weak) achievement. Scientists often express the hope that they will "continue"—that they will find challenging problems and projects. This is as true of older scientists as of younger ones. A forty-six-year-old physicist commented:

I hope I will continue to be able to do new things, learn new things. I think it's very important when you get to be more senior that you don't lose the ability to jump into something new [even when] you don't know whether you're going to succeed or not. You have to assume you will be able to learn the stuff or figure it out.[35]

How does she envision a conclusion?

There's no one big thing, I don't think. I guess it's the hope that I will be able to continue and also that I will be able to have some effect on younger people. . . . There might come a time when I decide to do something else, not research.[36]

A scientist at an earlier point in his career (age thirty-five) highlights similar themes.

I'm in a state now—I think I would like to do interesting work now, for the rest of my life, but I don't have any ultimate goal out there at this point that I'm trying to reach. I'm actually thinking about starting to do a little bit of science writing on the side, nonfiction articles for magazines.[37]

The image emphasizes steadiness rather than exponential growth or decline. Pluralists may embrace becoming, but they do so at a markedly slower pace as well as with a measure of hesitancy: both scientists quoted above comment on how their science commitments may turn in different directions at future points. Viewed against the backdrop of the moral career in science, these scientists have leveled off at intermediate phases. No longer is the drive for superior achievement intense. Their achievements by career's end will probably differ only modestly from their present achievements (which, of course, may be highly admirable).

[At the end of my career,] I don't know. I think there's a folklore that physics is a young person's game and that after a while you can't hack it anymore. Most of the older people end up going into administration or slowing down. I could very easily spend the rest of my life here . . . but I don't think I would be particularly unhappy to do that. I suspect that as I get older I will get involved in non–research activities to a greater extent.[38]

Notice that the imagery here of old age differs from that in the elites' accounts. For elites, being sixty was not considered a liability. By contrast, this scientist presents an image of slowing down and changing course. He invokes a stereotype to portray the social order of which he is a part— physics as a "young person's game."

Intraworld Differences: Off the Beaten Track

The unfolding of careers and of the self hinges on turning points (Hughes 1958) that result not only in changes in objective status but also in changes in individuals' subjective evaluations of their progress. Judging themselves according to the norms of their work environments, people may find themselves behind or ahead of schedule or completely offtrack. Since no world is perfectly uniform, let us turn to several elites, pluralists, and communitarians who have deviated from their local norms.

Elites

Elites stressed persistent activity from the beginning of the career to its end. Scientists often explained the persistence of their efforts by the levels of their scientific interests: they say they work hard because they enjoy it—tensions, frustrations, and hardships notwithstanding. But even in a world that stresses and rewards advancement, and a world where one daily sees others attempting to go forward, not everyone advances at the same rate, nor do all members remain faithful to this overriding institutional goal.

Elites who decelerate illustrate several common themes. They are typically around their mid-fifties. Normally they have been credited with major scientific achievements. Thus, they are not people who could be said to have grown discouraged as a result of early failures. In quite the reverse situation, many are adjusting to early success. They ask themselves what they should do next, and they set themselves apart by deciding to ease off climbing more steps.

Such scientists often reevaluated their professional priorities. They developed new interests: teaching was newly emphasized. Many found themselves backing away from research and showing greater interest in interaction with students. Along this line, many showed a more pronounced "expressive orientation" to their work. They grew less concerned about competitions and instead assigned greater value to the interpersonal dimensions of their relations with others. They grew apathetic about the judgments that colleagues might make of them. In this regard, they were independent-minded. They no longer held as their own goal the institutional goal of getting ahead. These themes arise in the account of "Charlie Hoffman."

> I think you can do very involved stuff professionally that is very high profile [and] very shallow in the sense that you put a lot of effort into it and it doesn't give you a lot of human return. I guess the bottom line is that no matter how accomplished I were as a physicist, after I retired there would be no interaction that would carry on [that is, like what I have] with my children. There would not be an interaction that is close, caring, rewarding. I don't see that in these professional matters. It's important to have them [professional matters, involvements]— there are important uses—but I don't think of it as a fundamental goal that you really strive toward with all your might at this point. I have done very good stuff. I know what it costs to do very good stuff, at least with my talents. It requires total absorption. Maybe if I were

Gell-Mann I could do it with my left hand while I was playing tennis with my right, but I'm not. For me to achieve really high level stuff, it just takes total absorption. Having done it and having seen how I felt after doing it, it's not that it wasn't worth doing, I just don't want to do it again.

Do you dream about anything now in terms of your career?

Not really. I'm very satisfied with what I'm doing. I don't know that I have any great aspirations. I don't lack ambition, but there's no driving goal that I'm obsessed with.

Is there something you want to accomplish or achieve?

I would like to live until my youngest son gets out of high school. That's my goal. I had a melanoma about six years ago, which I thought was going to kill me. When you have that kind of experience—it was clear that the most important thing to me was that I live until my kids were reasonably set, on their way. There wasn't much else that was important to me. Although I'm way past any probable recurrence of the melanoma, it is still an attitude that I carry around.

Do you have other things you would like to achieve or accomplish, that you dream about?

No, not particularly. I started out in West Texas, and my father had a seventh-grade education. Sometimes it's amazing to me that I became tenured at a major research institution. It's just an enormous distance from where I started. I went to high school in West Texas oil fields, where sports were the most important thing. I think that being where I am is an enormous, although serendipitous, distance and achievement. I don't think of things that I would like to achieve that are really driving me.

How do you envision yourself at the very end of your career?

About the way I am now. I was looking out the window the other day [while I was] at some meeting that was kind of boring and thinking, gee I wish I could live forever. My kids are beyond the point where you feel you have a lot of input. My youngest is fifteen, and my oldest is eighteen. By that point you certainly have input, but it's not the same anymore. So that worry is gone from my mind in terms of obsessing about it. My career is pretty comfortable right now. I think I've got a lot of respect as a teacher, and that's important to me.

How would you like to be remembered by colleagues?

By colleagues? I don't care. That's not important to me.

How would you like to be remembered by others in general?

As a good father—I think that's important to me. As a decent human being—that's important to me. I have that reputation.

What about your life do you think will outlive you?

My children. I've taught a lot of people. I've put a lot into being accessible and that sort of thing. That's not a very concrete thing. . . . Students go away. They essentially disappear. It's sort of amorphous. It's not a continuing relationship. It's more of a state of mind that you have in relating to individual people. Especially dealing with freshman, it's hard not to feel some generalized father thing. It's very hard not to think of them as my children in some sense . . . to help their self-esteem, to make them feel valued, to motivate them. Eighteen and nineteen—they are very young. They are just leaving home. It's a very uncertain time, when you are trying to establish yourself. Just by virtue of my age and my position, I become an authority figure before I open my mouth. It really is a lot of influence. If you use it in a reasonable way, for their best interests, I think you can make a lot of difference.[39]

This scientist's adaptations likely stem from his altered physical health. In other scientists's accounts, we see variations on the theme of self-reevaluation. In contrast to the dominant pattern of elites, some switch paths. Again, these patterns are evident in the careers of scientists who have reached relatively advanced phases. As the account below illustrates, their attention is focused outside of research.

I'm really dreaming about more time for things like . . . more time for outdoors stuff. I'm wondering about retirement more. I get hassled by a zillion things pressing me now. I'm teaching two courses, plus things on this [space] project are behind schedule. Why am I spending all weekend staring at this sophomore stuff, then the junior stuff? I'm sixty-three now; that's not what I was telling myself I was going to be doing. I have a dream of finishing this textbook and seeing that used successfully. I'm really puzzled as to what my dream should be after this [next space] launch: dive in and get in there elbow to elbow with the thirty-year-olds? And do the science? Or say: "Great, guys, have fun with it. We gave it to you, and I'm going to do something different now"?[40]

Still other scientists decelerate differently. They remain active in their research, but they approach pursuits with a realization that they will never likely yield a major breakthrough (though they may contribute to basic understanding in other ways). As the following sixty-three-year-old scientist illustrates, they see their efforts in smaller, more pragmatic terms, as opposed to the larger, grander ones that characterize the central pattern.

What do you dream about in terms of your career?

Now, at this moment? I still have a lot of nice ideas. I want to work them out, see them come to fruition. I don't think of any great [ideas]; I'm too old to think I can have revolutions in physics. But I'm doing a lot of nice things. I'm doing some of my best work now.[41]

Scientists also deviate from the moral career by accelerating. Among elites, the life course is socially controlled so that even early achievements do not excuse individuals from continued effort. (Considerably more leeway exists in the pluralist and communitarian worlds, which are more apt to include scientists who embrace alternatives after making a mark.) Indeed, achievements that come "ahead of schedule" raise people's expectations (for the achiever and others alike) because comparatively greater time is available to reach more luminous heights. Scientists who fill such shoes are given names that testify to their notable passage: "stars," "heroes," "hotshots," "early bloomers," "rate-busters," "path-breakers," "pioneers," "boy wonders," "golden-haired boys" (see Kanter 1977).

For most scientists, acceleration is consistent with what the moral career prescribes: continued effort. (The exceptions have been noted above.) Scientists who accelerate may be the beneficiaries of early success, but they also carry the burden of moving on to greater things. They advance to another phase of striving, and although they may never actually achieve greater reaches, their efforts are consonant with what their world mandates.

At the age of thirty-six, "Professor Wu" became a full professor in the physics department where he now works. Several turning points marked his rapid rise in the field, foremost among them the publication of seminal articles that resulted in key theoretic contributions. A prefatory statement on his resume reads:

> Professor [Wu] joined the Physics department at ——— in 1982 after approximately ten years with the Theoretical Physics department at ——— — Laboratories. He has made key contributions to the theory of disordered electronic systems and is a pioneer in "mesoscopic physics," the study of small devices at low temperatures. His current research interests include transport through quantum dots, fractional quantum Hall effect, and the theory of high temperature superconductors.

At age forty-five, nine years after arriving at the university, Wu was elected first to the National Academy of Sciences and then to the American Academy of Arts and Sciences, two of the highest scientific honors. The preceding year he was appointed to an endowed professorship after being offered a position by the University of Chicago. All such distinctions are devices of social control. Recognition regulates the manner in which the life course is supposed to be lived and experienced. It directs, and often inspires, effort. Consistent with this view, we should not be surprised that Professor Wu, conforming to the institutional goals of science, emphasizes striving even with praised performance.

> There is still a lot to accomplish in terms of the research; there's still many unsolved problems. It's an unending story. My hope is to be able to stay in this game.[42]

A similar orientation to work is evident in a scientist whose early achievements in space physics brought him international and popular renown.

> [At the end of my career I envision myself] probably hobbling to a meeting and renewing acquaintances between former students and colleagues. I don't think I'll ever quit actually. I think I'll keep going. I just get something done and do something else. I don't think I would ever really retire from work.[43]

These two scientists have caught the occupational disease of elites; they conform to moral mandates that obligate their continued investment in science. Achieving one great thing results only in the realization that something greater must now be accomplished.

Pluralists

Outlooks take on considerably more varied forms among pluralists. To bring the variation among pluralists to light, compare the following accounts.

> I'm involved with undergraduates at an introductory level—and I keep telling them that this course was designed so that you cannot go outside the building or do anything without recognizing the physics involved in what you are doing. . . . What's important to me is to have a satisfaction of understanding. Therefore I will continue to have a satisfaction of understanding based on my ability to understand. And if I don't publish it, it doesn't bother me. In fact I've never enjoyed publishing. Understanding is the important thing. Getting down to the nitty gritty and getting into a journal is not how I would like to spend my time. I guess the answer is that I don't have the constraint, the pressure, to make sure that the whole world knows that I understand something, and therefore I will spend my time understanding.

How would you like to be remembered by colleagues?

As a person who knew what he was doing. I guess the thing that grates me the worst is if somebody questions whether I am competent at what I am doing.[44]

Earlier in this interview I had asked, "Is there some ultimate thing you would like to achieve?"

No, not as a physicist. I'm very struck [by] something my wife commented to me on, and that is the comment by Jackie Onassis: if your children don't turn out well, it's all for nothing. That is actually my philosophy in life. It is very important that if you have children that you see that they reach their capabilities. [T]o me, my own person, seeing that my children are successful is more important than anything I can think of, no question about it. Because ultimately you cannot be successful; ultimately whatever you do, you end up dying. And that's it. They can probably bury your Nobel medal with you, but beyond that, what good is it going to do you when you're dead? So far, I feel very happy that I've done what I've done. I think about decisions before I make them, and I'm very happy with the decisions that I've made.[45]

In this account we recognize themes more often apparent in communitarian narratives (and in the fraction of elites who decelerated): an emphasis on teaching; an expressive or interpersonal orientation to work; the diminution of research productivity and recognition; explicit mention of the value assigned to familial roles. Notice as well that this scientist cannot assume a predominantly teaching orientation among pluralists without experiencing some strain (which is not necessarily true among communitarians). His withdrawal from research has been accompanied by implicit (and at times even explicit) questions about his competence.

I get angry if someone in the physics department complains about anything I do. Occasionally things will happen, "Why didn't you do this, why didn't you do that?" The answer is: I didn't do this because I was busy. I have been involved in women's rights. Someone can scream at me for twenty-four hours nonstop and it would not faze me in the least. I'm absolutely unfazeable, so I just ignore that.[46]

Thus, although pluralists tolerate variety, they are by no means free from the tension that stems directly from their variety. Tension is perhaps most apparent between scientists who decelerate and administrators who eternally seek to boost institutional prestige.

Other pluralists resemble elites. The theme of striving is as apparent in these pluralists as in elites.

How do you envision yourself at the very end of your career?

Vigorous in science, and continuing the mentoring role. I have a pleasure of collaborating closely with several very fine people, one of which is a professor of chemistry at another university, who at age sixty-eight runs a research group of twelve people and keeps on producing materials that we collaborate on. So I don't have a role model of declining at age sixty and going into semi-retirement mode. He still runs around the world giving his invited talks and is still writing his research grants. I'll probably do the same thing.[47]

When I look back, I can't believe that I was as fortunate and as successful as I was in terms of fulfilling what I wanted to do. Every time I set a goal, I achieved it. I was always driven by goals. I was a very goal-oriented person. When I was an undergraduate, I wanted to get into a good graduate school; when I was a graduate student, I wanted to get a good postdoc; when I was a postdoc, I wanted to become a faculty member; and then once I became a faculty member, I wanted to compete for national awards; and when I won one of those, I wanted early tenure. Now that I have early tenure, I guess I want to become a full professor. I guess the thing that I'm most concerned about now is that I would like to remain an active member of this astrophysics [community]. I want to continue to do good work. I don't want to back off and not take chances anymore because I have security. I would like to be in the National Academy someday. That shows respect of your peers. I don't have a goal of winning a Nobel Prize because I think it's accidental if you do research that garners the Nobel Prize. I could have already done it and not known it. I have some role models of people I would like to be like when I get to be in my fifties and sixties, which right now [at the age of thirty-six] seems like an incredibly long time away. One of the guys who I work with here . . . is unbelievably active. He's in his fifties, and that's how I'd like to be. So I do pick people out and say that's how I want to be in twenty-five years. And I pick people out and say that's how I don't want to be in twenty-five years.

Who are those people?

They are people who don't do anything anymore and just rest on what they did twenty years ago, and in this field, nobody cares about what you did five years ago, much less twenty.[48]

Clearly, some pluralists embrace the moral career in much the same way as do most elites. Continuity of action marks their outlooks. They strive for achievement and for recognition by honors, awards, and so on that encourage and regulate performance. This is evident even when careers accelerate: succeeding in a school career; being rewarded with favorable posts; winning national awards and early tenure. In many ways, therefore, some pluralists speak the language of elites.

Communitarians

Like elites and pluralists, communitarians are not wholly uniform in their outlooks on the future. The dominant career patterns consist of "getting stuck" and of foreseeing a displacement of interest from science to outside entrepreneurial attractions. Scientists who depart from these local norms fall into two patterns. In one, scientists carve out a research career along the outlines of the elite mold, resulting in a miniature version of the moral career. Because of the structural and cultural constraints imposed by the communitarian world (see table 2), however, this version cannot possess all the desired characteristics. In the second pattern, scientists withdraw from research altogether and embrace teaching, or they became communitarians as dedicated teachers in the first place, without ever entertaining high hopes, or strong interests, on the research front.

The outlooks on the future of those communitarians who show commitment to research appear little different from the outlooks of some pluralists and elites.

How do you envision yourself at the end of your career?
I don't really plan to end it. For example, I may retire as a university professor, but that doesn't mean that I'm going to retire as a physicist. I will continue to work on things that interest me. I want to continue to try to solve some of these little fine points that haven't been worked out and tie up some loose ends. I've got some writing I want to do. [I've] been working on three textbook contracts now for several years, and because of getting diverted off onto other things, [I] simply haven't completed those, so I'm probably going to finish those. In general I would basically like to stay active.[49]

The ultimate goal is to be more and more involved in research and get more and more positions of responsibility—being a leader, leading in many different things. I'm a leader in my group. We have the American Physical Society, presidents and secretaries and vice-presidents and all these things. I am a member of some committees. But I would like to be in a higher position sometime, making big decisions, making an impact on research. So [I would like to] contribute not only

from a research point of view but also sitting on important committees, chairing some committees where big decisions are being made.
I'm already starting to get in the channel. I would like to make a significant contribution to the field. I also would like, and I'm trying now to
do that, to leave a good impact on my students.[50]

I want to continue doing research and publishing papers. To me, publishing a paper dealing with a pedagogical subject is just as important
as publishing something in *Physical Review Letters*. One of the things
that I did find very fulfilling was going to China. I've been to main-
land China four times, giving talks and seminars at various universities. I found that very fulfilling, very satisfying. And I think going to
Brazil—I had collaborators at the Institute for Theoretical Physics
there—so I think doing some traveling and lecturing, being active in
research as long as I can.[51]

By invoking imagery of persistence, these scientists illustrate the manner in which they have led, or are attempting to lead, careers that parallel
those of elites. Some communitarians, exclusively younger ones, may
even view their present role as an intermediate step before taking a position in a more conducive research environment.

In some sense I regard this still as an intermediate spot. This is not necessarily where I want to be for the next thirty years as far as what I'm
doing and my level of contribution to society. The teaching is worthwhile, and the research is certainly satisfying, except that I think perhaps the scope or magnitude of the problems that I'm addressing
right now is not what I would ideally like to address. It all comes back
to this thing where I would ultimately like to be able to say that, yes, I
helped solve some problem that we face today in our society through
a technological innovation or development.[52]

I have been approached by other universities . . . and those were better universities than this one . . . the University of Tennessee in
Knoxville, and I would have had a dual appointment with Oak Ridge
National Lab. I thought about it, but I just decided that the move
would not have been good at this time. I certainly would have had
other possibilities [for research]; it's very flexible. But this university
has been very good to me. I feel that here I'm getting really good treatment. I already had my students and got my DOE [Department of En-
ergy] funding and got some start-up funds from here and started
building up my research. Interrupting that would have hurt my research.[53]

Scientists who fall into the second pattern view themselves primarily as teachers. They view research, when they do it themselves, as a kind of hobby, something not to be taken so seriously as to cause them to always want to do more.

> My career is near the end. I see myself being here for maybe three more years. It sure would be nice if I could ever put together the stuff that I have been doodling with all this time—it's what I call duality. It has some relation to the field; some aspects are fairly well-known, [but] I know little more about that than most people. Most people don't care. But I would certainly like to be able to tie that together more tightly. If I had some time and some insights, it would be nice to put out a little monograph—whether a hundred people look at it, alright—but a little monograph, just hang it up for me. It's sort of like completing puzzles.
>
> *Is there some ultimate thing you would like to achieve?*
>
> That for me would be an ultimate. I am participating in an experiment to change how we teach our nonscience introductory course.
>
> *How do you envision yourself at the end of your career?*
>
> I envision myself as still doing my mathematical physics puzzles. Actually my wife and I have talked about putting in for a Fulbright to go back to Brazil to see how things are back there [having been there early in his career]. Some travel, and I hope to keep healthy.[54]

Like others who fall into this pattern of serving as "master teacher," he appears content. These scientists proved the easiest to interview, in large part because they personify characteristics they claim go into being outstanding teachers: helpfulness, empathy, friendship, support, and encouragement. They do not seem determined to get ahead; they appear quite stable. Their aspirations center on serving others, and to this end, they imagine their futures as a time in which to look back over a complete life rather than forward to an unending, never-to-be-fully-surmounted ladder of scientific climbing.

Power of Place, Power of Individuals

For each group of scientists and for every individual among them, we have seen what matters most in distinctively shaping self-identity: the people with whom one interacts on an enduring, face-to-face basis. People are a constitutive part of a place, whether that place is a university department, a law firm, a basketball team, a household, a city, or a nation. The characteristics of people endow a place with the power, in greater or lesser degrees, that compels individuals to adopt certain moral standards and expectations of performance in the roles they perform. Simply put, places have cultures that constrain and shape their members.

As we have seen also, other components of places come into play in shaping individuals: physical resources, such as buildings, equipment, and facilities; and monetary resources, such as funds for creative inquiry, personnel, and the physical resources themselves. In other words, places have structures that help establish the moral conditions of role performance while also delineating a set of opportunities and constraints within which people operate.

But whether we single out parts of places or take them as a whole, the power of place over individuals is substantial. Following Emile Durkheim, places are ultimately greater than the sum of their parts: their strength transcends individual will, and they endure as individuals leave or die and are replaced by others. In both this chapter and the preceding one, the unfolding of self-identity has occurred as individuals have adjusted their aspirations in ways consonant with local opportunities. When individuals enter the group of elites (or pluralists or communitarians), they are prone to act and perform like other members of that group.

We can, however, raise the following question: can the uniformities and the differences among worlds be explained by the choices that institutions make about what kinds of faculty members to recruit and by the choices that individuals make about where to work? Put differently, what role do social selection and self-selection play in contributing to the observed patterns? In effect, this line of thought questions the "power of place" and instead underscores the power that individuals exert on their own fate and on those of others.

Pushing People and Their Narratives Apart

It is useful to examine scientists who began their careers at roughly the same starting points and who then diverged. In doing so, we deal with prime evidence that suggests how contexts change people's outlooks on self and career.

One such starting point is the institution where scientists received their Ph.D.'s and the time when they received their degrees. In many ways the place and the time of Ph.D. constitute an appropriate means of assessing change in individuals because these factors provide an objective common denominator. Roughly speaking, students of like institutions undergo many of the same socialization processes, which include internalizing norms about expectations for future careers (Zuckerman 1977). Although there may well be individual differences in talent, ability, and motivation among those in a Ph.D. cohort, we can assume that the range of variation is relatively small, especially among those coming from the most selective graduate programs. Most people who enter doctoral programs are accepted on the basis that they have true potential to succeed as researchers.

The scientist quoted below received his Ph.D. in 1987 from an elite department of physics, and he now works in a pluralist department. He described his outlooks as a graduate student and then as a pluralist.

[When you were in graduate school,] did you have a sense of what you wanted to attain?

To have an impact on the field, to be respected and known.
Did you have a sense of where you wanted to stand in the field?
Sure. Again, to be well respected, to be considered a leader.
How did you come to arrive at this university?
Limited number of job offers when the time came.[55]

In a passage quoted previously, this same scientist accounted for an evolution in his perspective.

My aspirations were to be at a higher-prestige university, initially—
Berkeley, Harvard, M.I.T.—types of places like I went to as a
student. . . . Ideally, I would have, at that time, would have wanted to
be a successful scientist at one of those leading institutions.
How has that affected you?
My view of life and career and institutions in general has changed a
lot. I think I'm quite content not to be there. It's hard enough at what I
would consider a second-tier institution like [this one]. Given my
choice right now where to go, I would go to an institution like [this
one].
*How do you think the atmosphere is different here than you would find at the
places you mentioned?*
A lower tension level here—a lower feeling that you have to produce
from day one and also less feeling like you have to compete with your
peers in terms of stature.[56]

An evolution in outlook and in self-understanding has taken place for this scientist and may well continue to take place, as it characteristically does for his pluralist colleagues who find themselves adapting to their environment.

"Geoff Silverman" similarly received his Ph.D. from an elite department at roughly the same time (1985) as the scientist quoted above. But he

ended up entering the elite world rather than the pluralist one. Notice
how his perspective on self and career differs.

> I'm afraid my ambitions have grown. . . . You replace one anxiety with
> another. First of all I had [a] survival anxiety. Now I have the anxiety
> that I'm given this remarkable opportunity very few people in history
> have had, that is, to work for some major research university with a
> secure position. I have no excuse really. If I don't do something really
> good now, it's because I wasn't smart enough. . . . [The institution]
> provides a supportive environment and trusts me to do what I
> want. . . . It [the place] has really benefited it [the career] because the
> standards are just uniformly high and I'm not convinced that young
> scientists have terribly well defined intrinsic standards. I think the
> very best may, but generally speaking, I think people's research qual-
> ity and attention to detail is to a substantial extent colored by the ex-
> pectations of their colleagues . . . my colleagues are just uniformly
> high-caliber people—people with really high standards of research
> and integrity and commitment to the [professional] community. And
> because of that, I know that's how my parameters got set.[57]

Note how this scientist's ambitions grow whereas the ambitions of the
previously quoted scientist subside. Changes that result from the contexts in which scientists work are as close as we will come to a naturally
occurring experiment. If we had the ability to place Silverman from the
start in the pluralist world alongside the scientist who spoke first, we in
all probability would hear a different voice speaking of different visions.
Thus, even scientists coming from comparable graduate institutions at
comparable times come to see themselves in systematically different
lights.

The time and the place where scientists received their Ph.D.'s are not
the only means to assess how contexts shape individuals. Ambition itself
represents a means to measure the power of place. Many scientists had
quite similar ambitions when starting out, even those who received their
Ph.D.'s at different types of institutions. In large measure, I have explained such similarities by the heroic myths that tempt all people who
embark on a career. Ambition for greatness may be more pronounced
among those who receive their training at elite schools simply because
elites have the closest brushes with greatness and more often mandate
greatness as a personal goal. In other words, there is generalized ambition in elite institutions. But in science, as in many other endeavors, universities basking in prestige are not the only places to find people who
have ambition. Romantic visions of the ideal career often know no

boundaries because the myths of heroic accomplishment themselves permeate the wider culture.

The content of people's career narratives should be affirmed, for this is where "contextual effects" are also evident. Throughout the preceding chapter and throughout this chapter, unfolding—change in the career and in an individual's subjective perspectives on the career—has served as the guiding idea. As I have shown, cooling-out—the process of realigning one's achievement expectations with local opportunities—is the chief mechanism associated with this unfolding. In the end, cooling-out leaves individuals more or less content with the available chances of success. Accordingly, it occurs as a contextually based process by which self-identities change: cooling-out produces different results in the different worlds of science.

The Labor Market

Social selection and self-selection, and their potential impact, must also be viewed in historical context. Academic jobs have been scarce in the late twentieth century. As a result, the possibility of individuals self-selecting institutions for employment is considerably slim. Currently, those who might have some "choice" about where to work more often than not constitute a small group indeed. "Choosing the right pond" becomes more a mental than a practical exercise, if it is an exercise people put themselves through at all (Frank 1985). As documented in chapter 1, many people who enter Ph.D. programs in physics never receive the degree, and of those who do, nearly half do not get physics jobs (whether in the academy or outside it).

It is possible that some individuals self-select by eliminating institutions from consideration, places such as prestigious schools where they believe they stand a poor chance of receiving offers or for which they have ill regard. It is unclear how widespread this phenomenon might be. On the one hand, since competition for jobs is high, eliminating possibilities from the start would sharply reduce one's pool of opportunities. Alternatively, some people (such as those without elite degrees) may take a more realistic view and concentrate their efforts on places where they feel they have the best chances. However exercised, self-selection was likely more widespread in times when individuals had more choices for where to work.

In perhaps larger ways, prevailing market conditions have fostered social selection—the wide latitude that institutions of all kinds have in hiring candidates. Whether market conditions have been tight or loose, institutions have always had control over the people they hire. But when conditions are tight, schools have even greater control over the people they allow to enter their ranks. Given the magnitude of competition, the candidates selected by departments are likely just as capable as many of the people turned away. This is a key component of "institutional upgrading," wherein schools are in a market position to employ people of a higher standard compared with those employed in years past. In principle, all three types of institutions benefit from the surplus of academic labor by selecting the most outstanding candidates. But pluralist and communitarian schools likely realize the greatest gains, since they now hire people who once might have entertained thoughts of employment only among elites. In addition, it should be pointed out that although departments have this discretion, they may not always elect to hire graduates of elite departments.

Thus, social selection and self-selection play active roles in forming the worlds of science (and also all other occupational worlds), but impact is in part a function of the market. Even when selection comes into play, we have seen how contexts are apt to affect individuals. The elite, pluralist, and communitarian worlds remain distinct in fundamental ways, not simply because they select individuals who are distinct to begin with but also because they induce systematic change in the individuals who enter and make passages through them.

No (Wo)Man Is an Island

The ups and downs of careers cannot be explained by individual factors alone. Like individuals, careers are socially situated. They unfold in interaction with an environment, as can be gleaned from the many accounts that stress the changes that occurred after people settled into their worlds. Even those hired because they were "promising"—as is nearly always the case—sometimes later see themselves, and are sometimes seen by others, as "disappointments." Whether an individual process or social one, the selection was made on the basis of promise, not "dry aspiration." In these and other ways, selection cannot explain the whole story of people's turning, twisting careers.

The contextual perspective that I have offered views the social environment "as a constituent force in development" (Dannefer 1984b, 847) rather than merely as a setting for preprogrammed maturation. The many individuals coming to terms with blunted ambition are as indicative of the power that places exert on people as are the many individuals who, on entering more facilitative environments, see their ambitions grow. In every instance, we witness changes, which largely come about through social interaction within different worlds that offer different things to the people journeying through them.

Feeling the Effects of Time

From the beginning, I have used time as a social dimension along which to understand how people experience and present their careers. I have defined time by cohorts, as done in many sociological studies (among the many examples, Clausen 1993; Elder 1974). Cohorts are groups of individuals who experience shared events. Those events in turn become markers against which people evaluate their past and future. Here, "the event" is the year in which scientists received their Ph.D.'s. In comparing individuals within and across cohorts, we see how individuals age in similar and in different ways. The resulting picture of the life course properly takes into account variation in how lives are patterned (cf. Levinson 1978, 1996).

Up to now, I have used cohorts to locate individuals at different points along the life course: in rough terms, those who are older, middle-aged, and relatively young. I have emphasized how people at these points make meaning of their careers, which are correspondingly long, medium-length, or relatively short-lived. Put differently, time has surfaced in how scientists of different ages view their unfolding careers and how they craft self-identities in light of this unfolding (Hughes 1958; Neugarten 1979a; Neugarten and Datan 1973; Wells and Stryker 1988). Simply through differences in age, scientists, like all other people, are exposed to different experiences and turning points in their professional (and personal) worlds, shaping them in distinctive ways. Moreover, the cumulative effects that experiences and turning points sometimes exert also are variously manifested in scientists who differ, quite simply, in how long they have been around.

I will use cohorts, though, in an additional way: to assess the impact of historical change on individuals and their careers (Clausen 1993; Elder 1974, 1979, 1981, 1991; Kanter 1989; Riley, Foner, and Waring 1988). Scientists of different ages account for their careers in different ways not only because they are exposed to different turning points and experiences in their interpersonal worlds but also because they are exposed to events that, occurring at unique historical junctures, have left a defining imprint (Mannheim 1952; Ryder 1965).

Social Change and the Patterns of History

I have suggested that the contraction of the academic labor market has increasingly pushed newly minted Ph.D.'s at elite institutions to jobs at pluralist and communitarian schools. I have used this historical trend to partly account for the occasional elite tenor among communitarians. This trend also partly explains why pluralists (as represented in this specific sample) bear more similarity to elites. In these ways, we see the role that social change plays in organizing a profession.[58]

The vicissitudes of the academic labor market, like any other labor market, are part of more general economic forces in society (Easterlin 1980; Fuchs 1983). The patterning of human lives—including people's life chances, social development, financial security, and personal satisfaction—is closely linked with the state of the economy (Dahrendorf 1979; Elder 1974). Particularly over the past thirty years, several major studies have adopted a life course perspective to examine the explicit linkages between societal economic conditions and individual life outcomes (Clausen 1993; Elder 1974, 1995; Elder and Conger forthcoming; Elder, George, and Shanahan 1996; Elder, Rudkin, and Conger 1994). Just as the conditions of one's birth shape not only one's first days but also subsequent years and often one's entire life (Clausen 1993; Elder 1974), the conditions that prevail when people start out in science frame their experience both at the starting gate and potentially all along the track.

An important point is easily lost. The first place of employment in science careers is crucial: it sets a horizon of expectation. Conditions prevailing when scientists start out in their careers situate their perspectives on the future. The significance of the first job is easily lost because it is peculiar in many ways to academic careers. Few other occupations are set up so that the first place of employment strictly delimits what one later does or where one later moves.[59]

In navigating the labor market, scientists are filtered to various types of institutions that support scientific enterprise in different ways. They come upon different opportunities that, for better or for worse, specify the road they will most readily travel. Resources, whether human, physical, or monetary, are unequal in availability and unequal in quality among worlds, but they prescribe in meaningful ways the kind of career one has. As we have heard, many scientists who once entertained dreams of major scientific discovery now find themselves modifying those dreams in accord with local opportunities (and newly perceived individual ability). Moreover, in entering a world, scientists are exposed to and internalize norms about performance, but as we have seen, those norms differ significantly from one world to another.

Furthermore, once at an institution, a scientist will often find it difficult to move to another, especially to another of a different world. As table 4 showed, scientists tend to stay at the places they first enter; mobility is low. These mobility patterns further highlight the significance of the transition from being a student or postdoctoral researcher to being a full-fledged member of academia. This transition sets the stage for the whole career.

A tracking system develops. The tracks not only are set clearly apart from one another but also are rarely crossed. The conditions under which careers begin are therefore key to understanding how careers unfold. The effect of time is only reinforced by the simple fact that where people begin their careers will, for most, be the place where they conclude their careers. In short, the hand of history ultimately affects the measure of one's impact on the world (Shanahan, Elder, and Miech 1997).

In times of fiscal austerity, careers mature under conditions radically different from those surrounding careers in times of relative prosperity. One is led to ask whether the "best" careers of today, in a time of relative scarcity, have the potential to resemble the best careers of days past, in times of relative abundance.[60] To use the cliché, times have changed. With those changes come different careers and thus different bases for crafting self-identity.

Unemployment

A primary indicator of the conditions affecting individuals and their careers is unemployment. In studying unemployment, we can ascertain the context in which careers begin to unfold and yield outcomes, which may differ significantly from the resulting outcomes had the careers begun at different times.

Unemployment rates compiled by the National Science Board, the scientific research arm of the federal government, present snapshots of socioeconomic conditions at different historical episodes. Table 8 lists unemployment rates for scientists in selected years between 1968 and 1993, the first and the last years for which such rates are available. A story line of the labor market trend in academia emerges from these rates, as follows. In the cold war and *Sputnik* eras of the 1950s and 1960s, jobs in physics and throughout much of higher education were relatively plentiful. This is reflected in a comparatively low rate of unemployment for sci-

Table 8
Unemployment Rates of U.S. Scientists
in Selected Years, 1968–1993

Year	Rate
1968	0.5%
1970	1.0%
1971	1.5%
1973	1.3%
1975	1.1%
1977	1.3%
1978	1.1%
1979	1.0%
1981	1.6%
1983	2.5%
1986	1.5%
1991	2.0%
1993	2.6%

Sources: National Science Board (1979, 1981, 1983, 1985, 1987, 1993, 1996).

Note: Unemployment rates are estimates. The rates may not be perfectly comparable across years due to different categories of compiling and reporting data. Rates for 1993, 1991, 1986, 1983, and 1981 are for "physical scientists." Rates for 1979 and 1978 are for "doctoral scientists and engineers." Rates for 1977, 1975, 1973, 1971, 1970, and 1968 are for "doctoral scientists." In all years for which data are available, aggregate rates are *lower* than rates for physical scientists and for physicists specifically. Thus the table *underreports* the rates for physicists. The years shown are those for which data are available.

entists in 1968, at 0.5 percent. Rates of unemployment for scientists were not even gathered systematically before 1968, presumably because there was no need for such information. As a 1973 government report stated, "For most of the 1960's the production of new scientists and engineers could not match the demand for their services" (National Science Board 1973, 59).

The U.S. system of higher education expanded throughout the 1960s. Anecdotal evidence suggests that people entering the academy during this time normally received several offers for positions, sometimes as many as a half dozen or even more. Tenure and promotion policies at many universities were not formalized. In many instances, faculty either never knew they were "up" for tenure or never bothered to be concerned because jobs were more or less plentiful and people could move easily.

The market contracted in the early 1970s, however, and jobs were not as plentiful as they had been throughout the 1960s. In 1971, the unemployment rate of scientists peaked at 1.5 percent, a figure that perhaps belies the gravity of the change felt by scientists who began their careers at this time. The market was tight. Several years later, it "softened." By 1979, the unemployment rate for scientists lowered to 1.0 percent. As the expression commonly used among contemporary graduate students goes, people were once again "getting jobs."

Then a major turning point occurred. Beginning at the onset of the 1980s, the academic labor market contracted sharply. This period is often characterized as one of the most dismal in academia. "Retrenchment" is the term often used to refer to this time in American higher education. Unemployment among scientists rose to 2.5 percent in 1983, the highest rate in history.

This situation lasted throughout much of the 1980s, with minor fluctuation from year to year in market conditions—but with no sustained period of recovery akin to what had occurred in the aftermath of the contraction in the early 1970s. A tight market situation persisted into the 1990s.[61] In 1993, the unemployment rate for physical scientists was 2.6 percent. For physicists specifically, it was 4 percent (National Science Board 1996).[62] The unemployment rate for physicists at this time was among the highest of unemployment rates for all academic fields; only one combined field had a higher rate (sociology/anthropology, at 4.6 percent), and only one set of fields had the same rate (geosciences) (National Science Board 1996). Currently, students and faculty members characterize the market as "extremely tight."

To summarize, science experienced expansion in the 1950s and 1960s, modest contraction in the early 1970s, rebirth in the mid-1970s, and a sharp contraction in the early 1980s that has more or less held up to the present time. Unemployment rates, though, tell only part of the story about the conditions that people face in their professions. What is worse, because unemployment rates for scientists today appear deceptively low, it may be difficult for these rates alone to convey the impact that different times have on individuals and their careers. Of what consequence is an increase of 0.5 percent in unemployment over a two-year period? There is perhaps a temptation to think the change trifling.

Unemployment in science is of course a part of wider developments, including the growth and contraction of the educational system as a whole. For some readers—especially academics and older ones at that—rewalking the steps of history in U.S. education will only confirm what is widely known. But it is important to put major historical trends on the table, in clear and full view, because they provide a fuller context in which to assess their impact on individual lives.

Enrollments, Faculties, and Money

Major historical shifts in American higher education are evident in three key areas of the system: enrollment, faculty size, and research expenditures. Table 9 documents the trends in higher education on each of these counts. Notice that in each of the components, the sheer numbers increase. In looking at those numbers alone, one might conclude that U.S. higher education has grown steadily and that, at any one point, it deserves a clean bill of health. Yet these numbers are not as significant as the *rates of change*. Whether we examine enrollment, faculty size, or research expenditures, rates of change capture major historical disruptions.

On each count, the 1960s appear to have been a golden age in American higher education. In that decade, enrollment increased by nearly 120 percent; the number of full-time faculty members increased by nearly 100 percent; and expenditures on research increased by well over 200 percent. These trends stem from a number of social, economic, and political factors. Universities began educating the baby-boom generation (Modell 1989) and hired an unprecedented number of teachers to meet the demand. U.S. universities began to compete internationally with vigor, in part as the result of political concerns stimulated by the cold war and *Sputnik*. This fostered growth in faculty size and a massive infusion of funds for research.[63]

In the 1970s, enrollment, faculty size, and research expenditures continued to increase in absolute terms, but their rate of increase was drastically lower: severe contractions were felt in all three of these areas. The rate of change in enrollment was 44.5 percent, and the rate of change in faculty size was 20.6 percent. Federal expenditures for research experienced even more drastic changes, creeping to a near standstill: less than 10 percent for basic research and less than 20 percent for academic research and development.

Even more intense contractions were felt in enrollment and faculty size in the 1980s. These components had a rate of change of 17.0 percent and 16.4 percent, respectively. Interestingly, contractions tempered in federal expenditures on research, with rates of change for basic research and for academic research and development hovering around 50 percent, up significantly from the previous decade. Nevertheless, those rates represented a severe divergence from the rates of the 1960s.

Systematic Differences in Career Beginnings

These trends historically situate individuals with distinct sets of opportunities and constraints. They potentially alter trajectories by closing

Table 9
Selected Trends in American Higher Education, 1949–1990

Enrollment[a]		
Year	Number[b]	% Change
1949–50	2,659,021	—
1959–60	3,639,847	+36.9
1969–70	8,004,660	+119.9
1979–80	11,569,899	+44.5
1989–90	13,538,560	+17.0

Faculty Size[c]		
Year	Number[d]	% Change
1960[e]	163,656	—
1969[f]	326,000[g]	+99.2
1970	369,000[g]	—
1979	445,000[g]	+20.6
1980	450,000[g]	—
1989	524,000[g]	+16.4

Federal Expenditures for Academic Basic Research[h]		
Year	Amount[i]	% Change
1960	1,659	—
1969	5,216	+214.4
1970	5,191	—
1979	5,594	I 7.8
1980	5,717	—
1989	9,047	+58.2

Federal Expenditures for Academic Research and Development (R&D)[h]		
Year	Amount[i]	% Change
1960	1,552	—
1969	4,878	+214.3
1970	4,760	—
1979	5,561	+16.8
1980	5,805	—
1989	8,307	+43.1

[a]Source: U.S. Department of Education (1995).

[b]Includes students enrolled in all higher education institutions.

[c]Source: U.S. Department of Education (1970, 1995)

[d]Includes full time instructional staff with the title of professor, associate professor, assistant professor, instructor, lecturer, assisting professor, adjunct professor, or interim professor (or equivalent); excludes graduate students who serve on instructional staff.

[e]Actual year is 1959–60 but has been changed for comparability with subsequent data.

[f]Actual year is 1969–70 but has been changed for comparability with subsequent data.

[g]U.S. Department of Education survey estimate.

[h]Source: National Science Board (1996).

[i]In millions and in 1987 constant dollars.

doors on people who might have had other doors opened had they entered science at another time—possibly as few as five years earlier or later.

Accounts of how careers got going highlight the impact of time. The following situation was common among older scientists—those in the first cohort—who entered the field in the period of expansion.

How did you come to arrive here?

I began to look for jobs with my mentor, and I began to look at a number of high-quality schools. My wife wanted to stay on the East Coast to be near the family, which was limiting. I didn't want that. Anyway, that was one of life's compromises. . . . I looked at Columbia, I looked at Harvard . . . I looked at M.I.T. and then, just sort of the residue of the past, I looked at Brandeis. I really didn't like Columbia, and I didn't even look at Yale; I didn't like their reputation.

So you were in a position really at the time to kind of choose or select?

Yes, I think I could have gotten other jobs, but I never tried. . . . So it ended up that they invited me here to give a talk, and then they immediately suggested I be on the research staff. So I laughed. I said if they didn't have a faculty position for me I would stay at Princeton. So they gave me a faculty position.[64]

The tightening of the labor market in the early 1970s is apparent in the next account, by a scientist who received his Ph.D. in 1970. Despite the hardship of the time, he found his way to an elite institution, a fact that marks him as exceptional.

Are there . . . major branches, forks in the road you've met?

Yes . . . [w]hen I got out, when I was looking for a job. In the early seventies, in 1970 . . . the first job crisis began. . . . [I]n the sixties the universities were building up rapidly, and there was a sudden crash, and that happened exactly when I got out. . . . You would hear stories of people who were not getting jobs, whereas up until that time there were plenty.

How did the market affect you?

I was offered an assistant professorship at Pittsburgh, which is quite unusual. There were very few jobs in those days. And most places require a postdoc first. But I turned it down and instead took an instructor position, like a postdoc position, at Yale.

Why did you turn the assistant professorship down?

First of all, I was stupid, I was naive. But really the reason was, I thought if I went to Pittsburgh, which is not a top-rank university, I thought I would probably be stuck there for the rest of my life. I thought maybe

I should take a chance and try to go to a more prestigious place and see. Miraculously it worked out for me. Looking back, you realize it was a tremendous risk.[65]

There is a clear contrast between this account and the preceding one. The fact that a scientist could more or less pick an institution—with the selection, moreover, limited by a spouse's wish "to be near the family"—must have seemed like a fairy tale several years later. But an even greater contrast is seen in accounts from scientists who came on the scene in the 1980s, when the labor market experienced its most severe contractions. The scientist below received his Ph.D. in 1986.

I spent a year and a half to two years in the process of finding [a job]. . . . I must have applied to fifteen or twenty different jobs. I had about seven or eight interviews. For a time I felt like I was always having great interviews, one- [or] two-day visits at universities and doing very well, and then getting a phone call a few months later and [hearing] that I was their number-two choice. . . . I had a three-year postdoc at Michigan State, and I had another one in Germany. It would have been possible for me to apply for this level of job after the first postdoc. I made the decision then that perhaps I would not be as strong a candidate for the job I eventually wanted with that level of experience.[66]

This scientist also illustrates the growing tendency of scientists in this period to use the postdoc position as a "holding pattern." His and other accounts point to the trend of taking not just one but two and sometimes even three postdoctoral positions. Scientists then expect to compete successfully for faculty positions, or they decide to enter nonacademic science or ultimately to leave science altogether. Professional societies' surveys of graduates bear out this pattern (American Institute of Physics 1995; Kirby and Czujko 1993).

Accounts of the times during which scientists began their careers are not particularly useful for conveying the frustration (or pleasure) prevailing in a period, though that is a real part of many people's experiences. Rather, like the information reported in tables 8 and 9, these accounts document the more general conditions that people face, the conditions that frame and shape them and their careers.

Degree Production

While higher education contracted through the 1970s and 1980s—downsizing in unprecedented fashion—people continued to earn doctoral de-

Table 10
Doctoral Degrees Conferred by American
Universities, 1949–1990

Year	Number	% Change
1949–50	6,420	—
1959–60	9,829	+53.1
1969–70	29,866	+203.9
1979–80	32,615	+9.2
1989–90	35,720	+9.5

Source: U.S. Department of Education (1995).

grees at a steady rate. Table 10 traces the pattern of doctoral degrees conferred by U.S. universities through this transitional period.

The number of people receiving doctoral degrees in the 1960s responded to the great demand for their services. Between 1960 and 1970, production increased by an amazing rate: 203.9 percent. This rate of change was not sustained over the subsequent decades of the 1970s and 1980s (which would clearly have caused a problem), but the number hovered in the same range, roughly between 30,000 and 35,000.

Intense competition has ensued. Intense competition characterizes the prevailing climate of U.S. science and all of higher education. Intense competition also describes an individual characteristic that many scientists increasingly appear to possess, if only because job scarcity elevates their concerns about being successful.[67]

Cohort Comparisons

A logical point at which to begin tracing the effects of events on cohorts is by considering how "success" gets defined and who succeeds, for success is what results from the institutionalized drive among scientists (and other professionals) to make significant contributions to a field and to life generally (Merton 1973c).

Standards of success in science change for straightforward reasons. For example, austere socioeconomic conditions raise expectations of role performance. Under tight market conditions, rewarded performance is raised to a higher standard because the labor pool is large relative to available positions. Competition among numerous individuals raises the threshold of what passes for success. At one time, a single nice article may have served a scientist well; at another time, two, four, or twenty articles may need to be published.

This applies not only to checkpoints at the beginning of careers, such as who gets a job in the first place, but also to standards throughout careers, simply because competition pushes all thresholds higher.[68] As is evident in many of the scientists' narratives, achieving one thing is often an occasion to achieve something greater. Those who fall short are easily replaced if they lack job security. But even when jobs are secure, individuals are easily outperformed because the pool of talent is so concentrated. Consequences are felt in comparatively less recognition, including both formal recognition, such as citations, honors, prizes, and salaries, and informal recognition, such as respect and esteem. One observer of change in the academic life put it bluntly:

> My first book was well received; I published some articles and review essays, and after three years . . . I was promoted to tenure without even knowing I was being considered for it. When I think of the recurrent ritual humiliations to which assistant professors are now periodically subjected . . . I wonder why any intellectual of independent mind would seek an academic career; it is hard for me to imagine anyone getting through it without lasting wounds, deep bitterness, and a taste for revenge. (Berger 1990, 158)

The changing standards for evaluating role performance, and the consequences for individuals and the groups to which they belong, are captured in table 11. This table lists several conventional measures of success and identifies how the different cohorts compare in each category. Here we see the impact of secular trends on individuals, trends discussed previously in terms of enrollment, faculty size, and research expenditures.

In fundamental ways, life as a university professor now is different from what it was just twenty years ago. And if we go back farther, the changes are even more stark. On every count in table 11, "success" grows more difficult to attain as the years go by. Today, even entering the profession takes a body of substantial contributions, reflecting just how competitive academics have become. This is evident in the average number of papers scientists had published by the time they were appointed as assistant professors, usually the first long-term academic rank held. Scientists who received their degrees before 1970 had published an average of 5.2 papers, an accomplishment in its own right. But by 1980, the number had changed drastically: scientists receiving their degrees after this point had published an average of 12.5 papers before getting their *first* academic job.

A similar pattern holds in the average number of papers that scientists had published by the time they received tenure. For the oldest cohort of scientists, this average was 14.6. But for the youngest cohort, the average was 23.5.

Table 11
Characteristics of Role Performance, by Cohort

Characteristic	Cohort I (Ph.D. Pre–1970)	Cohort II (Ph.D. 1970–1980)	Cohort III (Ph.D. Post–1980)
Avg. no. of papers at first job[a,b]	5.2	7.2	12.5
Avg. no. of papers at tenure[c,d]	14.6	24.0	23.5[e]
Avg. no. of papers at full prof.[f,g]	22.6	47.2	—
Avg. time to tenure (in years)[h,i]	3.9	4.5	4.8
% who held postdoc[j,k]	54.5	68.8	94.5
% who held >1 postdoc[l,k]	1.0	3.8	63.6

[a]Average number of journal articles published by the time scientists were appointed to their first job. Other publications are excluded. "First job" is defined as appointment as assistant professor or visiting assistant professor.

[b]Cohort I: N = 20; Cohort II: N = 15; Cohort III: N = 22. Cases do not agree with total cohort numbers (see table 1) because of unavailable data. Scientists who began their careers outside of academic science (e.g., as industrial scientists) are excluded.

[c]Average number of journal articles published by the time scientists received tenure. Other publications are excluded.

[d]Cohort I: N = 20; Cohort II: N = 14; Cohort III: N = 10. Cases do not agree with total cohort numbers (see table 1) because of unavailable data or because scientists do not have tenure. Scientists who began their careers outside of academic science (e.g., as industrial scientists) are excluded.

[e]This number is *underestimated*; it includes only two cases of elites and thus overrepresents pluralist and communitarian scientists, who tend to be less productive in research.

[f]Average number of journal articles published by the time scientists were promoted to full professor. Other publications are excluded.

[g]Cohort I: N = 18; Cohort II: N = 11; Cohort III: N = 0. Cases do not agree with total cohort numbers (see table 1) because of unavailable data or because scientists are not full professors. Scientists who began their careers outside of academic science (e.g., as industrial scientists) are excluded.

[h]Average length in years it took scientists to receive tenure.

[i]Cohort I: N = 18; Cohort II: N = 11; Cohort III: N = 0. Cases do not agree with total cohort numbers (see table 1) because of unavailable data or because scientists do not have tenure. Scientists who began their careers outside of academic science (e.g., as industrial scientists) are excluded.

[j]Percentage of scientists who held postdoctoral positions or equivalent before becoming assistant professors or visiting assistant professors.

[k]Cohort I: N = 22; Cohort II: N = 16; Cohort III: N = 22.

[l]Percentage of scientists who held more than one postdoctoral position or equivalent before becoming assistant professors or visiting assistant professors.

Two points are in order. First, the average number of papers published by the time of tenure for Cohort III (23.5) is actually underestimated, because only two elites are included in the count—they are the only elites in the third cohort who have tenure. Consequently, the measure overrepresents pluralists and communitarians, who tend to be less prolific. Thus, if more elites were included—as they would be at a later point in time—the average number would be far higher and would represent a marked difference from the averages for Cohort II (24.0) and Cohort I (14.6).

Second, it is instructive to compare numbers *between* each of the success measures. As pointed out, the average number of papers published by members of Cohort III when they got their first job was 12.5. Notice how close this average comes to the average number of papers published by Cohort I when they received tenure: 14.6. Had members of Cohort I entered the profession some twenty-five years later, and done the same work, they most likely would not have been retained by the institutions where they were working. Similar contrasts are evident between the number of papers published at time of tenure, which under normal circumstances is earned after four to seven years, and the number of papers published at time of promotion to full professor, usually taken to be the pinnacle academic rank. For the oldest cohort, the average number of papers published by the time they advanced to full professor was 22.6. But members of the youngest cohort had published an average of 23.5 papers by the time they received tenure. Even as an underestimated measure, it exceeds what older scientists had done to achieve the highest rank.

The average time it has taken scientists to earn tenure also has changed, lengthening over time. Evaluation of early role performance now occurs over a more protracted period. In a tight market, universities have the luxury of keeping individuals longer at the ranks of assistant professor and associate professor. The oldest scientists received tenure after an average of 3.9 years, whereas the youngest scientists received tenure after an average of 4.8 years.[69] Although the difference is less than a year, the changes in other aspects of performance make this difference more profound. On average, young scientists are far more productive than their older counterparts, and their productivity occurs within an only slightly longer period of time.

The remaining two measures—dealing with postdoctoral positions—also indicate secular change in the profession. Slightly more than half of the oldest scientists (54.5 percent) took a postdoctoral position, whereas nearly all of the youngest scientists (94.5 percent) assumed such a position. The postdoc is now nearly universally considered a necessary career stage. It has become institutionalized, and the exceptions are now those who never take this appointment.[70]

The institutionalization of the postdoc likely stems from a variety of factors. On the one hand, cognitive developments in the sciences have rendered this a crucial stage in which to acquire further specialized expertise if one is to successfully conduct research at later points in the career. On the other hand, tightened market conditions have elevated the postdoc as a career stage in which to gain competitive advantage for academic and nonacademic positions. This competitive advantage is not distilled by the fact that nearly all scientists now take a postdoc; rather, the advantage is gained through other standard performance characteristics acquired in this stage: the quality and quantity of publications; the reputation and influence of faculty sponsors; and the prestige of the host institution. The scientists' accounts characterize graduate school as a basic training stage; the postdoc makes or breaks the career.[71]

> Essentially there is no graduate student who directly goes to a faculty position. . . . As a graduate student, you work with your advisors. You should demonstrate your ability to work independently, to come up with ideas by yourself. . . . [I]n your postdoc, you demonstrate your ability to initiate new research. . . . The very important thing is the hot topic you can get—if your work becomes influential and . . . people know you. . . . If you are good, you will get a faculty position after two years.[72]

The latter view of the postdoc—as a means of gaining competitive advantage—is, if anything, confirmed by the number of those who held more than one such position. Only 1.0 percent of members of Cohort I took more than one postdoc; the percent for Cohort III was 63.6, a remarkable change. These figures illustrate the dramatic effects of shifts in higher education. The contraction of the system, combined with steady doctoral production, created an oversupply of labor; a backlogged pool of individuals use postdocs to outcompete one another. For many, the postdoc is the sole means to keep their feet planted in science.

That observation raises a further point. Table 11 tells us only about those who have remained in science. Considering the increased rates of unemployment and the continued contraction in academia throughout the 1990s, we should view these individuals as a lucky few. We lack substantial information about those who had to leave simply because there was no room.

For those who do remain, however, table 11 is highly suggestive about the role that ambition plays in a career in science. Regardless of the time, ambition propels individuals forward, pushing them to persevere when the nature of the task is inherently difficult. But the role of ambition appears to have grown more intense as the profession itself has grown more competitive. The standards against which scientists are now formally judged are unmatched in historical comparison. Institutionalized stakes in science are as high as they have ever been. If we compare those stakes with other activities throughout society, science is likely one of the most competitive games in town. If ambition plays a part in people's success, science more than ever renders ambition a prime ingredient for a career of any consequence.

Outlines of Cohort Identity

As one might suspect, the distinctions of careers over time do not lead to an "age-irrelevant" community of scholars (Neugarten 1974, 1979b). Loosely speaking, "period effects" have been thought to define groups in distinctive ways as a result of the "imprints" left by historical times or events.[73] The shared conditions in which people come of age and live characterize the cohorts to which they belong and give rise to the idea of a generational identity (Mannheim 1952; Ryder 1965). Accordingly, popular parlance speaks of "generation gaps," discrepancies in outlook and personal conventions between age groups. A manifestation of these discrepancies is the idea of "generational politics"—behavior and policy clashes between age groups. Familiar imagery pits the young turks against the old guard, the Roundheads against the Cavaliers.

One of the most prominent aspects of the young-old divide among scientists concerns the performance expectations that groups have internalized. Often these expectations reflect the different historical conditions within which their careers began. Thus, the young complain about how older, senior colleagues fail to conform to the expectations they themselves use to judge the merits of junior scientists. This is especially true in the pluralist and communitarian worlds, where new hires often have elite blood and where senior faculty members continue to embrace, in their own work, the standards of days gone by. By extension, the old, particularly pluralists and communitarians, complain about how their institutions no longer sufficiently value teaching— which for many senior faculty has been a primary concern. Instead, the university is now "research-oriented" or, worse, "run like a business." The politics of age comes across in varieties of ways, as the accounts below illuminate.

> Physics is a young person's game. Especially with the new laws, which mean that there is no mandatory retirement for professors, we are already facing a bit of a problem. People are not retiring; they are not making room for younger people, and that's a problem. I'm director of a lab, so I have to deal with these things. I know that many . . . people believe in their hearts that they are better than any three assistant professors. But the fact is, they are not.[74]

The problem here is that people inside [the university] don't recognize [the importance of research] or are afraid of it. I think one of the things that has happened was that this was a teaching school at one time and people were hired to teach and now [that] research has become important, there is no way that they can make up that difference and become something they were never hired to be. I think they see a loss of control; I think they always envisioned that they would eventually be in charge and move up the ladders of the school, and I think there is a real fear . . . that value is placed on research. There's no way they can compete in that arena. It's almost like they have been passed over by a generation.[75]

Reprise

This discussion has focused on intercohort comparisons. Historical conditions have differentiated science careers across cohorts by directly affecting opportunities and constraints in work and by successively altering performance standards. In seeking to uncover the ways in which lives are multiply patterned, life course research in sociology has stressed the importance of making not only intercohort but also intracohort comparisons—seeing how subgroups *within* cohorts differ in systematic ways (Clausen 1993; Elder 1974; Elder and O'Rand 1995).

In this work, intracohort distinctions are largely manifest in contextual effects—in the changes that worlds of work bring about in people. Within any one cohort, careers in elite, pluralist, and communitarian worlds play out differently, in systematic fashions. I have described the principal social mechanism involved in producing these changes as "cooling-out." Whether speaking of the older, younger, or middle cohort, we see patterned differences in how careers unfold as a function of context.

In similar ways, contextual distinctions extend to the idea of "cohort identity" presented above. I have discussed the ways in which cohorts of scientists, in any one world of work, come to see themselves and their careers in terms of being/becoming and acceleration/deceleration. Contextual differences represent the different ways in which members of single cohorts age and the different ways in which they see themselves.

The professional life course takes many forms. It is not uniform, a series of monolithic stages, standards, and transitions that apply equally and evenly to all. Rather, the life courses of individuals interact with, and are shaped by, macro events and forces. In very direct ways, scientists, like all other people, do not transcend history but are closely linked with it. The opportunities at hand, and the perspectives on the future that those opportunities engender, follow in direct ways from the historical period in which people find themselves. Contours of experience, measures of misfortune or prosperity, shapes of careers, and the extent of impact—all such aspects of an individual's self-definition arise and play out under the forces of different times.

Conclusion

Scientists form one of three styles of outlooks on their futures. These are best described as gradations along a continuum of being and becoming. Put another way, scientists' outlooks embrace varying estimations of mobility aspirations. The will to get ahead is strongest among elites. The elite world best facilitates, and its members mandate, an upward climb. Communitarians most often "get stuck." The will to succeed may be strong, but communitarians are often handicapped both by limited opportunities and by a culture that stresses service before self. Pluralists' outlooks are directed at achievements of a middle range.

In each world, however, people deviate from local norms. The grip of social control is never so tight among elites, or so loose among pluralists and communitarians, that it forces all individuals to conform to dominant patterns. In each setting, some individuals bear similarities to those in adjacent worlds.

The ways in which scientists create future selves have served our main purpose: to ascertain the role that ambition plays in careers. We have uncovered the forces that motivate, and that fail to motivate, people in their careers. In speaking of how they envision their futures, scientists have used the same imagery that they use to orient and fuel their actions.

Anticipation of status encompasses the overall scheme. The culture and structure of the elite is especially conducive to anticipatory status socialization because consistent and stringent social control regularizes the life course. Elites invariably remind one another that there is always something next. Ambition among pluralists and communitarians may not, on the whole, be so overweening because they possess fewer ways to control one another. The results are seen in their diversity: motivations are varied, with fewer incentives to persist. It is easier for individuals to reduce ambition, or to let go of it altogether, when others working around them have done the same. Thus we may conclude that given the forces at play, clones at the starting gate in each world would very likely look altogether different by the time they reach the finish line of their careers.

FIVE

Selves in Conflict and in Harmony

In chapters 3 and 4, I described the ways in which scientists accounted for their pasts and their futures in order to unite dimensions of the self to form a present, but continually evolving, professional self-identity. The conceptual themes running through each of the chapters centered on dreams and imagined selves: scientists presented self-images in light of subjective appraisals of their pasts and of their anticipated progress toward desired ends. We see ambition in their dreams and in their actual efforts to get ahead.

We need not view ambition exclusively through dreams and imagined identities, however. In dreams and in imagined identities, people cast themselves in favorable lights. As we have heard, they craft self-identities by imagining themselves in ways that often bear little relation to reality. Thus I propose in this chapter to study ambition from an additional vantage point, that of people's weaknesses. I will specifically refer to this as *self-doubt:* the insecurity that individuals feel about their status. In examining scientists' self-doubts, I seek a holistic view of individuals, complementing their dreams with their vulnerabilities.

I seek this holistic view for theoretic reasons: self-doubts tell us as much about what drives people (or holds them back) as do their dreams (Horney 1945; Ludwig 1995; Simonton 1994). Confidence and insecurity are not mutually exclusive, nor is it a contradiction to find both characteristics in a single individual. They may operate together. At times, confidence may succumb to insecurity; at other times, confidence may operate as a cover for self-doubt.

Insecurities may be manifested in two forces that drive individuals: a need to succeed, coupled with a fear of failure. We likely see ambition when individuals try to allay or rectify their perceived weaknesses.

[Woodrow] Wilson's own reflections of his youth furnish ample indication of his early fears that he was stupid, ugly, worthless, and unlovable. . . . It is perhaps to this core feeling of inadequacy, of a fundamental worthlessness which must be disproved, that the unappeasable quality of his need for affection, power, and achievement, and the compulsive quality of his striving for perfection, may be traced. For one of the ways in which human beings troubled with low estimations of themselves seek to obliterate their inner pain is through high achievement and the acquisition of power. (George and George 1956, 8, quoted in O'Brien 1994, 194)[1]

In what follows, we will hear the conversations that people have with themselves. In all their angst and contradiction, inner voices often speak a sentence not of death (except in the most troubled individuals) but of denial. We will hear these scientists navigate the fear and anxiety that at once threatens and creates their ambition.

Dealing with Demons: Inner Voices and Self-Confrontation

Academic worlds engender distinct psychosocial anxieties. Individuals encounter a range of career problems that impinge on the ways in which they see and understand themselves. Some of these problems arise more often in one world than in another, given the distinct cultures and structures. Other problems are present in all three worlds.

Elites

For elites, psychosocial anxieties typically revolve around progress. Elites often doubt their accomplishments, even though high and steady accomplishment is one of their customary features. Indeed, doubts about progress likely flow *from* high and steady accomplishment. Because achievement is morally mandated, scientists easily find themselves making interpersonal comparisons.

Once again, this phenomenon may be explained by social control. Institutional mandates and group norms compel effort and productivity. Doubts about progress emerge in light of scientists' efforts to conform to the moral career. Thus, we find elites asking themselves how well they measure up in any one phase. Elites are especially prone to ask, "Am I good enough? Will I succeed? Do I have what it takes? Am I doing as well as I should be? Am I slowing down? Will I keep up the effort?" And when careers fall off-track, elites are most apt to wonder why, simply because their culture conditions them to move onward and upward. Comparison is implicit in all of these questions: like other scientists, elites create their self-identities vis-à-vis others, and all turn to the moral career as their guide.

Mandates to achieve induce anxieties about people's very achievements and about the rates at which these are made. This applies to nearly all elites, even those who deliberately decelerated. Anxieties about progress know few boundaries in age; the young and the old doubt their accomplishments. In spite of how eminent some people become, they are always able to identify someone believed to be better.

"Charlie Hoffman," introduced in the previous chapter, explains the standards with which he and other scientists work and the anxieties that these standards induce. His account illustrates how someone who deliberately slowed his career nevertheless experiences doubt about his standing in the field. In spite of distancing himself from institutional goals, he cannot completely escape the control that those goals have over individuals.

> People in physics tend to compare themselves to people you can't compare with—Einstein or Gell-Mann or Feynman. In some sense that's the standard and even though it's a ridiculous standard you're always wondering whether you're a good physicist. So certainly I've had doubts about how good a physicist I was, but I don't know any other physicist who hasn't had those doubts.[2]

The following scientist talks about whether he doubts himself.

> Yes, daily. Daily. I don't have any doubts about my technical abilities. I have doubts about whether I will have another good idea. Will I ever come up with a new problem? My wife says that—we've known each other for nine years—she says it's perfectly clear: there is a cycle where at the low point I'm just miserable and I can't think what I'm ever going to do, and it may last a couple of days. Then some ideas will come up and other problems will begin to emerge, and then I will rise, and there will be this frenetic activity as things take shape and a paper emerges and concepts develop and everything builds. Out comes a paper, and then down [I go] again. She says it's absolutely clear, and she doesn't worry anymore because she has seen it and she knows with every trough there will be a peak. At the bottom there is this terrible fear that I will never think of anything ever worth doing, that I've published my last paper.[3]

The imperative to get ahead is felt especially when conditions appear adverse to that goal. Temporary setbacks are magnified into larger generalizations about the prospects of one's longer-term mobility chances.

Nobody has a career which is just uphill. Things don't work out; you get frustrated; nothing seems to be working. The experiments you are doing just aren't [working]; people aren't interested; you aren't getting interesting results; they are not going forward, etc., etc. There are dry periods. You think at those points that you've lost it. You begin to wonder if you are ever going to do anything good again.[4]

Eminence does not necessarily bring security or unchallenged confidence. A Nobelist commented on how the competitiveness of elites often inspires feelings of inadequacy.

If you just look around you, especially in a situation when you are trying to solve a hard problem and aren't making progress, and you see [others] who are successful, very subtle things they can reach—you have doubts.[5]

Similarly, a woman scientist who has risen to the top of her field commented on the perpetual role that self-doubts play in scientists' lives.

I always have doubts about myself. I'm always plunged into a situation that I feel totally unqualified, unprepared to do, and somehow you have to rise to the occasion. I say two things to myself, and that helps me overcome this awful feeling that I'm not qualified to do x,y,z: One, somebody else thought I was qualified, because otherwise they wouldn't have appointed me to do it; and second, in most cases this is the first time a woman has been asked to do this, and not only do I have to make myself qualified, but I have to do a good job.
What are the specific doubts that enter in?
Oh, that I don't have the background, that I don't have the skills, that I'm not smart enough, all of those things.[6]

Anxieties about progress result in an obsession with time: how "little" of it there is; and how it is most "efficiently" spent. Elites usually go to great lengths in guarding time (cf. Hochschild 1997; Nippert-Eng 1996). They commonly hold "office hours"—times at which they allow themselves to be seen and heard. Their visiting hours are normally set only at scheduled intervals during the week or exclusively by appointment. They seem to discourage random interruption, informal dialogue, and idle chitchat. They are intent on getting down to business. Colleagues who meet in the corridor after not having been in contact for a period are less likely to ask each other "How are you doing?" than "What are you up to these days?" (read: "How are you spending your time? Is it productive? Is it interesting? Will it lead to something worthwhile?").

A scientist conveyed common sentiments about time and how this usage is viewed in conjunction with one's hoped-for progress.

What would you like to be better at?
Better at being able to use my time more efficiently. If I have something that's in the back of my mind, I can't sit down and figure something else [out] very quickly. I just can't. I just think about that thing on and on while everything else is going to the dogs. I give my students something to work on; they come back; my mind is just not set on those things. And when I finally come to terms with what I was thinking about, then I can reconcile the rest. [It makes me feel] very sick. If you utilize time over several days, you feel very energetic at the end, you feel very healthy. But if you have just gone around in circles the whole day, making no progress, it's only frustration. It's a very sickening feeling.[7]

Psychosocial anxieties about progress, questions about ability, and obsessions with time intensify competition for recognition. Recognition validates one's efforts and ultimately the self, or so it seems. More than pluralists and communitarians, elites appear in regular need of reassurance about their self-worth. Often they do not feel sufficiently honored, in large part because honor is so central to their self-identities. They often appear so uptight and self-absorbed that these may be normal characteristics, part of a constellation of attributes that compose their "emotion-work" (Hochschild 1983): the conscious manipulation of impressions in order to appropriately perform their roles.

Scientific elites normally exhibit egoism and ego-strength: they take what they do (and often themselves) very seriously. They believe in the importance of their work and are inclined to think that those who aren't similarly motivated are clods or charlatans or are somehow not "with it." They are not easily persuaded, nor are they easily impressed by others. And they have been socialized to withstand the extremes of adversity, perhaps because they encounter relatively large doses of it in the competitive games they play.

Elites sometimes seem oppressive and imperialistic about their viewpoints and intellectual agendas, if only because having an agenda is a necessary part of being a leader. At times they are dismissive of alternative or competing ideas and the people who espouse those ideas. They often exhibit a need to be "in control." Communitarians and some pluralists might regard them as arrogant. But to each other, they appear driven by serious interest; many also see themselves as the guardians of high standards (cf. Parsons and Platt 1973). Given their standpoints, they normally learn to tolerate the intellectual abrasiveness and the austerity that are side effects of their seriousness.

Competition for recognition affects many personal attributes and mannerisms, including humor. Elites do have humor, but their humor often possesses distinctive features. Used at work, corny or witless humor would likely be met with a measure of neutrality or disgust by elites, for such humor, they think, corrupts the seriousness and professionalism that their world morally defines as key. In talking to elites, I gathered the impression that job behavior seen as excessively casual or cavalier carries the sentence of being "unprofessional," even incompetent. The person displaying such behavior appears shallow-brained and provincial and fails to bring honor to himself or herself or to others, thereby violating one of the elites' most sacrosanct moral codes. Elites are more likely to respond favorably to humor, including vulgarities, if it possesses intellectual or learned charm. Even this humor, however, is normally used only in the most informal settings. Moreover, elites are not prone to make fun of themselves, for that, they believe, is self-denigrating. Among close friends or colleagues, however, they jest about comic foibles. And their private humor off the professional stage may be as unrefined or unskilled as that found anywhere else. But even in fun, intellectual seriousness characterizes the overall scholastic orientation of elites.[8]

Many elites have grown so accustomed to success that it is expected in nearly everything they do. When success does not come, or does not come as quickly as they had expected, elites generally grow anxious and angry (sometimes at themselves but more often at the actions of others who have caused the disappointment). One scientist conveyed these traits by seeing in others many of ambition's ancillary characteristics.

> People here are typically ambitious people, overachievers. And they frequently feel that they have not achieved their full level; they feel that their colleagues are not respecting them. A lot of people want to command respect from their peers, from their colleagues.[9]

Notice how the scientist quoted below claims an elite school to be more humbling. He raises a paradox. On the one hand, he benefits from his institutional location in ways that most other scientists do not. On the other hand, he says his position is not very gratifying. In spite of the prestige a group might confer on individuals, the incapacity to be at the top leaves some feeling underacknowledged. Such social conditions likely heighten ego-insecurities among those whose ego-investments are typically large.

[This university] has a lot of people in each field [of physics] so that even within a place like [this one] it's harder to stand out. You've got a faculty of very high quality so that any one of us in a small school would be the best in their field, would be the outstanding person in the field in that school. [Here] you are just one of a half-dozen. So it's more humbling. It's not for everybody. This level of attention is not very gratifying. Maybe for me it's okay because I didn't expect to be at the top of anything when I started.[10]

He also underscored the need for recognition.

If you were a very special type, you could just stay by yourself and think and get no recognition at all and still have a good time. But it's hard without any positive feedback. There are very few individuals who can sustain that; they have to be strongly decoupled.[11]

When recognition fails to be won, the effects are often dispiriting. "Geoff Silverman," a scientist quoted in previous chapters, tells such a tale. Even though his success has been steady and swift (or *because* it has been steady and swift), rejection is difficult to digest.

I was quite disappointed by not getting a Sloan Foundation Fellowship. My disappointment was rather amplified by the fact that the Foundation then wrote to me and said "Will you apply again next year?"—which I did, and then they still didn't give me one. I was bruised by that. Every time I hear the Sloan Foundation I think, I don't like you. That upset me. There are probably other things like that, the occasional proposal that's turned down. But I content myself with the fact that most people have to put up with much more rejection than I do. I don't handle rejection very well.[12]

Silverman's story about a dear colleague points further to the anxieties that elites often feel about recognition.

I think that there are lots of really super people that don't realize that they are super. There are really excellent physicists who are still neurotic and anxious and worried about their image. And I'd like to put my arm around them and say it's alright, you've done well. I was just in Cambridge a few weeks ago, and I had lunch with a man called — ——, who is probably one of the two most powerful people in British science. He's the person who rescued my career. He's approaching sixty-five, and we had dinner, and he was regretting the fact that he . . . he was expressing the fact that he sees himself as being underappreciated. And to me this guy is the god; he's somebody who has clearly made it; he's done beautiful, wonderful things, and he's just an inch away from the Nobel Prize, which he will never get now, but he's done everything else. I think he's marvelous. I adore him. I think he's a really wonderful man. And he was really upset about the fact that people don't cite his work and don't know his papers, and so I would say to him, "Name a paper." He'd name one, and I'd say I know that paper: "1958, you did this and this, and that paper now forms the basis of this body of research," and he'd say, "Oh, I didn't know that." I could see that he was genuinely cheered by the fact that I knew about his work. So I think what people care about is that their work is appreciated. That's the most important thing. I must admit, every time there is a little message on my answer phone that says [Geoff], will you come and give a colloquium in Vancouver, or [Geoff], will you come and give a talk in San Diego, I think, Yes! One needs those constant prompts that people haven't forgotten you out there, that you are doing alright.[13]

Communitarians

Communitarians exhibit different psychosocial anxieties. The call to achieve and to keep achieving is not as strong; anxieties about progress are thus tempered. As is common in other worlds, when such anxieties do surface, they are typically in scientists who have yet to achieve tenure or promotion to a more advanced rank. But after tenure and formal promotions have fallen into place, communitarians exert little control over one another in the directions and rates at which their careers move. Scientists decelerate at comparatively early career phases (for example, at the onset of becoming a full professor) without suffering much in the way of negative sanctions or social stigmas. Many might consider this a plus.

Communitarians accept individuals who turn their attention to other endeavors, such as teaching and pedagogy, after early career phases, even if this comes at the expense of an active research program. In brief, communitarians ask one another to do well enough to advance among standard ranks, after which time they have considerable leeway in what they do and in where they go. As a result, communitarians' anxieties break into several patterns, with some having little to do with work.

Although personal problems may afflict elites as much as communitarians, communitarians are much more apt to discuss them. Why? Their structure of work is different. After communitarians achieve advanced statuses (typically full professor), they enter a terrain that allows them considerable latitude in the judgments they render on themselves. As long as they have advanced up the ranks, they are in an institutional position to assert self-harmony. This being so, nonwork problems enter their narratives more easily. In effect, more "space" exists to discuss extraprofessional concerns because often work no longer plays a dominant role. Thus the rhetoric that communitarians use to describe their self-identities more often draws on personal issues.

In the first pattern of communitarians' anxieties, scientists appear at ease with themselves professionally. Self-doubts about the professional aspects of their lives do not surface as salient issues. The following is an exchange with a communitarian scientist who has been a full professor for ten years.

What would you like to be better at?

I'd like to be better at tennis.

Has there been a significant time in which you felt things had not really worked out the way you wanted them to?

No. I think they have just about worked out the way I've wanted them to.

Do you have any regrets?

No, I don't think so. I'm a pretty happy guy, actually. I'm pretty pleased with myself. I'm confident about what I do. I have a lot of confidence in what I can do, not just in science and not just in teaching, but in other things too. I told you I wanted to be an architect. After coming here in '74 and buying one house and then moving into it, moving into another one, I started drawing plans for [yet another] house. And so I bought a piece of land outside the city . . . and built a house, from scratch.

Have there been times when you feel that you've let yourself down?

You mean done less that I thought I should?

Yes.

No.

. . . haven't lived up to your own standards or expectations?

No, I always do too much.

Has there been an inner conflict or personal turmoil that you have sought to understand in your life?

I can't think of one. An inner turmoil?

Yes, something of endurance that has been a source of concern.

Endurance.

An enduring source of concern or distraction.

Oh, something that I've carried throughout? I can't think of anything.

[Something] that you've tried to understand or resolve.

I can't think of one. Nothing comes to mind.

Have you ever tried something where you have been doubtful about whether you can do it?

Oh yes, sure. Building that first house was that way. But after building that first house, which was a 3,500-feet under-air, two-story, English tudor, then I built another house on the other side of town with another wife. And I built another house that was 4,000 square feet plus 1,400 in four garages. It's a big house. There's no problem with that kind of stuff. Now if I had any regrets, I guess I'd like to be smarter. I'd like to really be more intelligent. I'd like to retain more information. But I guess I'll just take the genes the way they are and do what I can with them.[14]

Earlier in the interview, I had asked about "success" and what it means and about his visions of the future. His responses underscore the relative harmony that communitarians achieve in early-to-mid career. The tone of the narrative emphasizes comfort, rather than strain, in how the scientist sees himself.

Is there some ultimate thing that you would like to achieve or attain?

Yes, I would like to have a house in Cozumel and in Colorado and flip back and forth between the two. In Colorado and Cozumel, Mexico. Have you ever heard of it? It's a great diving spot. The best diving in this part of the world, unless you want to go to Australia.

How do you envision yourself at the end of your career, whenever that may be?

I don't really see any reason to retire right now. I travel all the time. I get to do what I want to do. I go to Europe twice a year to conferences. I go to Colorado three times a year, sometimes four. Usually around Christmas I'll stay for almost two weeks. I'll end up going back there in the summer once or twice. So it's a pretty good deal. I get to do about what I want to do and still get ranked number one in the department.[15]

Thus, this first pattern stresses harmony; little psychosocial anxiety is apparent.

Communitarians that fall in the second pattern are also at ease with themselves professionally (even though aspects of their professional lives might be cause for alarm in other academic worlds), but they discuss anxieties that stem from their nonprofessional lives. They may have decelerated, but they have done so without experiencing tremendous social consequences. Typically their research has taken a backseat. Because pressures to perform at high levels are comparatively weak, these com-

munitarians capitalize on a freedom that allows them to subordinate discussion about professional anxieties (if they exist at all) to talk about nonprofessional anxieties. In the following account, I discuss self-doubt with
a scientist who has "stalled" at the rank of associate professor. In all likelihood, he will never be promoted.

What would you like to be better at?

I'd like to be a little better at playing hockey. I can't score goals worth a
diddly-squat. I'd like to be able to remember people better, remember
names and stuff like that. Really as far as who I am and what I am doing today, I guess there's nothing I would like to be able to do better. I
don't have any real strong features or characteristics. I'm not that
bright.

Have you had doubts about yourself?

You always have those because you always run up against something
. . . for example, whenever you grab something and run off with it and
say, "I'm right, I know I'm right, and certainly I'm right," and somebody goes, gee, you're wrong. That's a bit of an exercise in humility,
and you have a momentary doubt and say how did I ever come to that
conclusion. But you recover from that. I don't have any lingering, continuous, chronic doubts.

*Has there been some inner conflict or personal turmoil that you have sought to
understand in your life?*

I would like to be able to develop a relationship with another person, a
woman, that had any lasting power to it. None of my relationships last
very long. And I don't quite understand that. I've been married twice.
This is my second marriage, and that's coming apart at the seams.
Chances are that I'll be alone. I don't think that I'll be able to develop a
relationship with anybody that will last. I've known many women. I
don't have any friends either. As far as female relationships go, none of
them have lasted. I've had some intense ones, and I've had numerous
ones, but none of them have lasted very long. There's something
wrong with that. I have a defect in the machinery somehow.[16]

The third anxiety pattern that is characteristic of communitarians
highlights more professional themes. Yet though anxieties about work
are evident, they still differ from those characteristic of elites. Some
communitarians doubt their performance, but their doubts have less to
do with self-efficacy than with the purpose of science. They question the
value and meaning of research and are cynical about its usefulness. Such
scientists have not reached luminous heights and then suffered a "mean-

ing crisis," as was partly evident in some elites. Instead, they typically have had a modest research commitment.

What doubts do you have about yourself?

I doubt that I will ever accomplish anything significant in physics. I think that is becoming more and more probable as time progresses.

Does that trouble you?

It bothers me, to the extent that it would be nice to do. That's always a goal that everyone has. Yes, that is a bother.

Are there ways that you think you come to terms with that?

The way to deal with that one is to pick less significant problems, which is limiting in itself. Once you work on a less significant problem, then obviously the outcome is not going to be very exciting to the top tier of people in physics. So that's self-limiting, to pick a project that really doesn't have much interest, much significance.

Has there been some inner conflict or personal turmoil that you have sought to understand in your life?

I suppose in a sense that I guess everyone and myself in particular are worried about the significance of what one is doing and what's the point of doing what I'm doing: is there a point? If one is making "good progress," then you get affirmed by what everyone else thinks of your work, and so that validates it in some sense. I don't get quite the feedback that other people would perhaps get. So that has in fact led me to wonder, is this the kind of thing that I should be doing? Maybe I should be off taking out the garbage or something, you know. I teach the course, and I try to give as much as I can there. So I feel like I earn my living. I don't think I'm overpaid. The State of ——— gets more than their money's worth from me, I think. I'm not ripping anyone off. It seems like what I'm doing is an honorable thing, trying to find out what I can about nature, whatever that's worth. It may not mean anything, but what else can I do?[17]

The fourth and final pattern is evident among those communitarians who have miniature versions of elite careers. In predictable ways, such scientists often experience "elite anxieties"—those about progress. The scientist quoted below discusses an enduring anxiety about success and failure. Paralleling the accounts of elites, his account highlights the persistent notion of a ladder. He is not content with easing up on commitment when security in age has allowed him to do so. He is worried about how far he has gone and about his pace. Above all, he has deep-rooted doubts about whether he has done as well as he had hoped.

What major doubts have you had about yourself?

There was always a doubt about whether I would succeed, either personally or professionally.

Have you ever thought you had failed?

If I look back, there are certainly times when I could have done better either personally or professionally. . . . When I became a tenured full professor I could have stopped doing my research at that point and taught. I always considered research a part of what I wanted to do. To be successful in my mind, you have to do research, so there was a drive to be successful. But I think it's just one of my characteristics that when I see something that needs to be done, either professionally or otherwise, I tend to get right on it. I don't say, "Maybe I'll do it tomorrow" or "I can always wait to the next day." And I think that's probably served me well in my career, because it's so easy to lay things aside and do other things at hand. It's so easy to spend all your energies teaching and not put in any extra effort to keep your research program going. So this aspect of being driven has probably served me well.

Where do you think that comes from?

I suspect I had some fear of being a failure in life. I came from a poor family, an uneducated family. I always felt out of my element in the society in which I found myself, and so there was this fear of failure and looking back at each stage, I felt that I wouldn't succeed: I wouldn't succeed in getting a Ph.D.; I wouldn't succeed in getting this research done. And I think that has accounted for my driving myself hard.

Did you sense that fear throughout your career?

I think so. I'm not sure how aware I was at all stages, but I think whenever I reflected, I sensed that I had this fear. . . . There were no aspirations in my household that I would be successful in any way or that I could be successful. I was usually associating with people from higher social or economic groups, and I always felt—perhaps "inferior" is too strong a word—but that I didn't belong.

How is it that you've gotten to where you are, given your origins and the beliefs associated with failure?

I think there was always a conflict between driving myself and giving up. At times I was always fighting with myself, one-half of me saying "give up" and the other "push ahead." I think I always managed to continue going at several stages of my career, and I think a combination of just being able to persevere until opportunities came along resulted in my being successful. I've always had to contend with this fear of failure and feeling out of place, and that has certainly continued to this day. I think it has hurt me personally, but you do the best you can with your life, and I think I have done reasonably well.[18]

Like elites, this communitarian appears driven by insecurities, and he compensates through exceptional performance.

Pluralists

Pluralists exhibit psychosocial anxieties that collectively represent an elite-communitarian blend. On the one hand, some pluralists are similar to most communitarians. They are at ease, their selves in harmony, or they are anxious about personal, rather than professional, matters. On the other hand, some pluralists are similar to elites. They doubt progress, performance, and ability.

Notice in the following accounts how the scientists appear at ease professionally. They convey a sense of feeling unpressured, comfortable, and generally happy. There is very little sense of angst, turmoil, or doubt about how these scientists see themselves in relation to their work. Both scientists are senior full professors.

What doubts have you had about yourself?
Not a whole lot. I think most of my doubts over the years were in the early part of my career where I wasn't always sure I was really smart enough to have a research career, and I've had one. And in other areas I haven't really had significant doubts.[19]
What doubts have you had about yourself?
I see my failings, and they are a source not of deep depression but some days you feel more depressed than others, but by and large I'm a cheerful sort of person.
What are the failings that you see?
Not as diligent as I could be in my work as a physicist. In domestic responsibilities, not as willing perhaps to take additional jobs. It relates to responsibilities, all of these. I'm certainly more comfortable with myself now than I have been as [a] younger person: A, because I carry these things out a little better and I'm a little more responsible, and B, because it doesn't bother me as much if I'm not. It doesn't seem to matter quite as much as you get older. I wouldn't want to go back. I'm quite comfortable where I am.[20]

In contrast to these scientists, other pluralists exhibit elite-like doubts and anxieties. There is a sense of pressure to perform and to keep performing, underscoring once again the idea of a continual ladder to climb.

You have doubts all the time in the sense that you have to keep up to be a professional physicist. You have to produce. It's like being an artist. If you don't produce stuff, you don't receive any grants and your stock just goes down. You go through bad spells where you're not succeeding; things go wrong; it's your fault. You made a bad decision. Your design for the apparatus was lousy, things like that.[21]

I think a fear that everybody in this field lives with is that you're going to stop having ideas sometime. You become good at something—there is enough data there that you can keep reanalyzing. The danger comes, or the fear that most scientists have to overcome is that you have to get over the fear of trying something new. You have to take risks. You have to be willing to stick your neck out. A fear that I have is that I'll get complacent. I don't want that to happen. I'd like to be able to maintain a happy balance between being a senior citizen or a relatively mature scientist at a young age, but I don't want it to come at the expense that I just write review articles for encyclopedias or for major journals all the time.

Have you felt that you have had to try to overcome personal weaknesses or insecurities?

Yes, it's always been a struggle for me to make myself keep working. That's what the big battle is all the time. You have to be self-motivated. You have to find something that makes you want to work fifteen hours a day, six days a week or more for very little pay and for very little chance of success, and that's difficult sometimes. I think I've been successful at it, but it's not something that comes easily.[22]

The scientist quoted below discusses his self-doubts and the competitive contexts in which those self-doubts are most evident. Especially remarkable about this account is the physicist's ability to see in himself—and to articulate—doubts that he appears to have carried over several phases of his life. He traces his doubts to family influences, and he speaks both as a budding physicist and as a son whose father is an eminent social scientist at a major research institution.

What doubts do you have about yourself?

Capability. Scientific capability. I have a lot of doubts. It's part of my personality. That's always been a part of my personality.

Where do you think this came from?

Oh, I know a lot of it is that my father has a lot of the same traits. It comes from your environment.

In what ways did your father express these characteristics?

Doubting the worthiness of his work. He was an academic also. Feeling he wasn't working hard enough and not getting enough accomplished, even while many people around him were giving him strong affirmation. . . . My family academically was always very high-achieving. Both of my parents have Ph.D.'s, and my two oldest sisters got straight A's and always did very well. So that was expected. And I was the youngest, and I just felt from my parents that when I did do something very good academically—the best in the class or whatever—their attitude was "that's nice." I [infer] now that my reaction was "that's not good enough." "If I'm only best in my class, I should be best in the state." I've had episodes of depression going through my career, especially at major turning points, like going from graduate school to postdoc, and then finishing my postdoc and finding a permanent position. Both of those times I had fairly major episodes where I was pretty depressed about my role in physics and what I was going to accomplish.

Can you elaborate on what the source of the depression was?

Uncertainty. Self-doubt. I think the good quality that I do have is perseverance. I didn't give up. Also I think I tend to think less well of how I performed. Other people think better of how I performed than I do. My own view of myself is lower than other people's view of me. . . . I guess there was this big fear of what happens if you reach a stage where you don't do well, you're not at the top of the class. That was unexplored territory for me, and so it [made me] very nervous to think about that. As you get higher and higher into professions, it's harder and harder and harder to stay at the top. I guess the biggest thing that I wish had been done different[ly] is to have a sense of security so that if you're not at the top, you don't feel like a failure. There are a lot of very good people who aren't number one. At this university here I feel very comfortable. I don't think I'm top, but I'm not the bottom either, and that's okay. I still know in general I have done well to be here.[23]

Toward an Assessment of Overall Satisfaction

In light of how self-doubt surfaces in different ways in the elite, pluralist, and communitarian worlds, we can now determine how overall satisfaction might be characterized for each group. In terms of peak experiences, elites are in a position to feel more "highs." Opportunities for advancement and intellectual challenge are more plentiful. But at the same time, there is an insistence in elite culture to continually build upon past successes. Thus, although opportunities for advancement are plentiful, so are opportunities for doubt about progress, leaving individuals in a perpetual state of never feeling fully satisfied.

There are only two ways that individuals are excused from the expectation to produce. One is through retirement, and even then expectations for continued performance may persist. The burden of expectation may be felt from others (colleagues or kin) or from one's own internalized work ethic. Elites themselves described how their visions of retirement differed little from their current way of life. It is difficult for them to let go. The second way of being excused from expectations is through incompetence (but even then, retirement might be forthcoming or already in place). In a somewhat rare show of softheartedness, elites normally excuse the senile from the standards against which all are supposed to work.

Overall satisfaction seems highest among elites whose dedication to work is most complete and whose success is at its most recognized. But this must be only a qualified conclusion, since for every success there is a reminder of something next to do. This feeling that life, even a monumental life, is incomplete and is marked with disappointments in one's accomplishments is poignantly captured in the biography of S. Chandrasekhar, the Nobel Prize–winning astrophysicist. Chandrasekhar, viewed as one of the greatest scientists of the twentieth century, provided his account several years before his death in 1995 at the age of eighty-four.

> I have tried to achieve detachment—detachment from having students, writing proposals, and so on. Let others worry, run things. But I have a feeling of disappointment because the hope for contentment and a peaceful outlook on life as a result of pursuing a goal has remained unfulfilled. . . . I don't really have a sense of fulfillment. All I have done seems to be not very much. . . . Science at the present time is greatly associated with haste and the desire to be at the top. But my unhappiness or discontent is not due to that, I think. Perhaps it's because of the distortion, in some sense, of my life, of it's one-sidedness, of the consequent loneliness, and my inability to escape from it all. Even at this stage of my life, when I send in a paper for publication, I constantly worry whether it will be accepted. I ask myself, why should I? Why should I? Why shouldn't I be able to spend the rest of my life reading Shakespeare? . . . I felt when I was young that when one reaches the age of mid-forties or fifties and one is moderately successful, one would have a sense of personal security and assurance combined with some contentment. I certainly haven't found them. (Walli 1991, 305–306)

Conversely, communitarians are susceptible to experiencing the most "lows." This, however, is greatly dependent on the extent to and speed with which individuals adapt when opportunities fall short of expectations. In chapter 2 I discussed the scientists' sentiments toward the institutions where they work. Communitarians appeared the least happy, voicing complaints about politics and "institutionalized mediocrity." Additionally, chapter 2 explored the ways in which communitarians are "cooled-out," leaving many of them with a negative and cynical residue. Those communitarians who appear happiest are those who have embraced teaching, either from the very beginning of their careers or in midstream as a result of reordering their aspirations.

An understanding of overall satisfaction necessarily involves considering how work intersects with other aspects of life. One way to cast further light on overall satisfaction is by studying marital dissolution. Although the fact has never been documented empirically, academics are often regarded as having notoriously troublesome personal lives. The extent to which their personal complications resemble those of other high-level professionals is also not understood. One hypothesis, though, is that there are strong similarities given the investment of time, competitiveness, and independence required for professionals to reach esteemed levels of success.

Marital dissolution is an imperfect measure of satisfaction, since for some it may bear no relation to work. Moreover, divorce should not be seen merely as one event spurred by one cause, although in some instances that may be true. Divorce is, of course, much more complex, often the result of many factors that have a cumulative effect over time. In addition, the dissolution of marriage is not the only measure that informs an understanding of people's general happiness. It is, nonetheless, a key one. To think that marital dissolution has *no* relationship to work would be equally foolhardy.

The percentages of scientists who are or have been divorced are listed in table 12. Divorce is most common among elites. By one interpretation,

Table 12
Percentages of Scientists Who Are
or Have Been Divorced

Elites	17.4%
Pluralists	5.5%
Communitarians	15.8%

Note: Scientists provided their *current* marital status. Marital history is taken from the interviews whenever explicitly mentioned. If there is any bias, these percentages underreport divorce, but probably by a small margin if at all. The data do not specify how many *times* a scientist has been divorced.

this is expected, since elites are the most competitive about their work. Work is central to them, as passages quoted in this and the preceding chapters have exemplified. If there is failure at work, problems in other domains of life are likely to arise, for work operates as a kind of controlling or equilibrating mechanism. Moreover, it is difficult to maintain legitimacy as an elite by cultivating a glamorously happy family life while maintaining a level of professional mediocrity.

It is also plausible that the commitment and the competitiveness associated with elites make them difficult people to live with. Their level of commitment requires a high degree of independence. The stakes for which they are striving require a kind of single-mindedness. Their competitiveness may be as much a mechanism of defense against rivals or against personal insecurity as it is a condition for achieving levels of success that by definition become increasingly scarce. As William J. Goode (1978) has observed, "Performances above the average become increasingly difficult to improve with each upward step even if the reward is great" (Goode 1978, 67). No less significantly, elites *enjoy* their work. Consequently, in pursuit of success that becomes more difficult, more demanding, and more engrossing, some err by neglecting other aspects of life.

Chandrasekhar's account conveys a sense of the toll that a scientific life takes on others. Ambition may have its noble causes, but it also clearly has its costs. These costs may at times lead an individual to wonder, as they did for Chandrasekhar, whether the quest is completely justifiable.

> It's not at all clear to me whether the single-minded pursuit of science at the expense of other, personal aspects of one's life is justifiable. Not so much for oneself, but particularly for those with whom you are associated. The person who suffers the most is, of course, one's wife. Was one justified in imposing that kind of life on that other person? . . . I don't feel I have to do as much in science. In fact, if anything, my feeling at the present time is the opposite. I mean I've worked in science for over fifty years now, and I don't see that I can accomplish very much more or that I can add to my scientific contributions by anything that will substantially change my record, unless by accident something or other happens. So if I continue to do science it's largely for my personal pleasure, and also because I do not know what else to do. . . . I've got so used to a certain way of life, that it's difficult to change. . . . You can have your life go by. [A fulfilled] life is not necessarily one in which you pursue certain goals, there must be other things. (Walli 1991, 306–307)

The second-highest rate of divorce is among communitarians. Notice that their rate of divorce is only slightly below that of elites (15.8 percent compared with 17.4). Explanations for patterns of divorce among communitarians are less certain. Perhaps those communitarians who approximate an elite career develop some of the personal characteristics associated with such ambition: independence, competitiveness, commitment, compulsiveness, bullheadedness. Alternatively, some communitarian scientists may grow frustrated because their initial expectations (together with whatever romantic visions they may once have shared with their spouses) fail to be even remotely fulfilled. Divorce can serve as a way to bury past dreams along with the frustrations that unfulfilled dreams create for those who once gave those dreams their blessing.[24]

The remaining group is pluralists, who have increasingly come to represent a middle ground. Pluralists operate in a world that allows them to spread their bets. A scientist can approximate an elite, can even become an elite in a pluralist world, but this is not mandated. By the same token a pluralist, like a communitarian, can embrace teaching, leveling off in research at midcareer, without suffering stigma. Professional activities and commitments may be embraced to a degree, sometimes to a *high* degree, but not necessarily at the expense of family and leisure. Life does not hinge on one thing alone.

The only pluralist who divorced (5.5 percent) clearly gravitated toward elites. It is therefore not by chance that patterns of divorce among scientists break along the boundaries of these occupational worlds. Disappointments in the past and present are most prone to threaten communitarians. Ambition is most likely to thwart the lives of elites. The one divorced pluralist, now in his mid-thirties, casts further light on these processes:

I went through a divorce, and basically that was because I was so competitive. I lost sight of things. I put my personal life on the back burner. I directly attribute that as a cost of being in the field that I'm in, unfortunately. I think a lot of people who are in competitive areas of science have had the same thing happen to them. Because the stuff I was doing was important, I got to travel all over the world. I couldn't afford to take my wife with me, and I think that's hard on people, to see that you're not making very much money and to be left behind like that. To be a spouse of a person like me, I think, was a pretty difficult thing to do. That price I did pay. I think I'm a more well-rounded person because of the experiences I've had. I think if I was like I was five years ago, I would say that I had not matured scientifically very much. I think I would be too driven, and I think I'm softer now. I think that makes it easier for me to get along, for other people to get along with me.[25]

Women Scientists

In attempting to advance our understanding of the life course, and particularly how it acquires meaning in the occupational sphere, I have adopted time and place as social dimensions in which to ground the analysis. Since time and place socially situate individuals and groups (cf. Elder, Modell, and Parke 1993), they allow us to broadly examine how careers take shape and how people identify with these careers over the course of their lives. We thus gain a more accurate view of how lives are variously, rather than monolithically, lived (cf. Levinson 1978, 1996).

This concern with variation leads us to gender. Gender is a central topic for most studies of careers: it has the potential to reveal interesting and important similarities and differences between men and women. Its treatment here, however, must occur within fairly strict limits. Of the sixty scientists in the sample, only four are women—a number that represents a slight *oversampling* of all academic women physicists in the United States (see chapter 1). Physics, and the sciences in general, are practiced primarily by men. This is changing gradually in all fields, at different rates from field to field (Sonnert and Holton 1995).[26] Thus, for more women physicists' careers to be studied, more women need to enter the ranks.

In light of these limits, the discussion that follows should be read with an appropriate measure of caution. The discussion is not meant to give a final, definitive statement but rather to provide some *suggestive* insights about the scientific careers of women, including the ways in which these careers may be similar to and different from men's.[27]

Women in Time and Place

To place the four women scientists in context, we need to know some basic background information. To begin, table 13 lists the four women scientists by their cohort and by their world of work. A subsample of four cannot, of course, fill every "cell" that the time and place dimensions cre-

Table 13
Women Scientists, by Cohort and Academic World

	Cohort I (Ph.D. Pre–1970)	Cohort II (Ph.D. 1970–1980)	Cohort III (Ph.D. Post–1980)
Elite	×		×
Pluralist		×	
Communitarian			×

ate, but this particular subsample does include people from each of the three academic worlds and from each of the three cohorts.

Like most of the men scientists in the sample, the women took positions and have remained at the universities where I interviewed them.[28] In this sense, their mobility patterns are indistinguishable from those of the other scientists. In terms of their Ph.D. backgrounds, the women show some diversity, but a majority come from elite schools. Both elites received their Ph.D.'s from elite institutions, reflecting the norm. The elder received hers from the University of Chicago, the younger from M.I.T. The pluralist received her Ph.D. from Cornell, and the communitarian from Virginia. Although the majority of the women went to elite graduate schools, their job placement parallels that of men: those who earn degrees from what are considered highly ranked departments normally feed into elite and pluralist worlds (but also increasingly to the communitarian world—see the discussion in chapters 3 and 4); graduates of pluralist and communitarian institutions generally find employment in other pluralist or communitarian schools.

All of the women held postdoctoral appointments, which also reflects the norm. The elder elite took a postdoc (on a National Science Foundation Fellowship) after receiving her Ph.D. in the 1950s. Postdoctoral appointments were not as customary then as now; taking such a position then reflected especially on the candidate's superior performance and on the subsequent competitive advantage gained by those who took the postdocs. The elites and the pluralist held postdoctoral positions at elite universities. The communitarian took a postdoctoral appointment at a major research laboratory that is affiliated with an elite university.

All but the pluralist are married. None of the women have ever been divorced. Only the elites have children: the elder, four; the younger, two. Both had their children at early career stages; neither had earned tenure by the time of the birth of her youngest child. (At the time of the interview, the younger elite was still untenured.) This is not terribly surprising, since the women were then ages thirty-four and thirty-five, within what are conventionally understood to be prime childbearing years. Nevertheless, the introduction of children creates an added responsibility, if not a pressure, and does so at a time when work performance is particularly consequential.[29]

The picture being drawn seems fairly uncomplicated and uneventful. But of course no life is free of complications or decisive events. It is the apparent "uneventfulness" that we should look at more closely. If we assume that women fulfill domestic labor roles in greater measure than men (Brines 1994; Gerson 1985; Hochschild and Machung 1989), these patterns suggest that women who are successful have an extraordinary organizational capacity as well as a drive that does not normally flag under competing pressures.[30]

Especially in a period of heightened competition in science, women who take on many diverse roles and who are perhaps less than chronically judicious in organizing time likely pay professional costs (and family ones too). For instance, because the onset of motherhood often coincides with the critical early phases of science careers, women may be more prone to drop out of science. They may also be more likely to self-select types of work worlds that are most conducive to multiple role commitments (for example, communitarian and pluralist worlds).

In the former case—women who drop out—this sample, consisting of employed physicists, offers little basis to test the hypothesis.[31] In the latter case—women who self-select institutions—the sample evidence, albeit limited, rejects the hypothesis. The women with children are elites, and the elite world is probably the most demanding professionally.

Women and Ambition

The simple fact is that we are dealing with four women—including two with children—who are all successful by conventional notions. The evidence seems fairly convincing for the claim made earlier: women, more often than not, must live up to high organizational standards. A related argument states that in order to succeed in science (and in other lines of work as well), women need to be more ambitious than men, whose role commitments are more centered. Yet if great ambition is demanded of women, so it is of men: intensified competition and the institutionalized drive for recognition make ambition requisite for all (Merton 1973c).

The women's narratives provide insight into how they view themselves and their careers and how ambition figures into their lives. As with all accounts, the best understanding comes from detail. But this is even more true here. In discussing a topic as complex as gender, and particularly in dealing with such a small number of women, we need to study the accounts of these woman on a one-by-one basis. Furthermore, gender and science have developed and combined into an active sociological research area (J. Cole 1979; Cole and Zuckerman 1984, 1987; Fox 1995, 1996; Long and Fox 1995; Reskin 1978; Sonnert and Holton 1995; Zuckerman, Cole, and Bruer 1991), but research must almost invariably work with small numbers of women, rendering our in-depth look even more necessary. In this way, each account places the narrator in the fullest context possible and thus puts us in the best-possible position to make qualified judgments about manifestations of gender in science.

Although gender issues did not compose a formal part of my interview questions—since gender differences were not the primary motivation of this study—they arose in the context of other, related questions. Consequently, each woman speaks about herself as a woman scientist in ways that are most meaningful and significant to her. We will study the women's accounts on those terms and also for the ways in which they are suggestive of more general patterns that might apply to other women like them.

I begin with the eldest elite, who has four children. Sixty-four years old at the time I interviewed her, she is an influential leader in science. To say much more about her accomplishments (which are many and great) would unfortunately compromise her identity.[32] Of the four women, she had the most to say about her role as a woman of science, perhaps because she speaks with the most experience.

How did the relationship between doing science and being a parent unfold?
    My first child came when I was a postdoc. I would say the biggest thing was getting organized. Going from zero to one child is the hardest. One to multiple [children] is easier because you're already into the swing of things. I think I did better; the adjustment was less. When my children were small, I only did my job and family. I didn't do anything else, practically. That was it. It was very localized. I almost didn't travel. My career was a little different from the men because I had four children—that keeps you hopping. I was not very active on the outside. I was just working and taking care of my babies.
How did you do it?
    I had a good husband[33]—that was a big plus. And I had the same baby-sitter forever. We found a very good person when I moved here. I made the job attractive by giving a good salary, good working conditions, and she stayed with me until she retired. She was very happy with the job, and she got a good-enough salary. She got a higher salary than her husband. It was a way to send her kids to college. They couldn't have gone otherwise.
How long was she with you?
    Twenty-nine years.
Twenty-nine years!
    It's amazing. I made the job attractive. My children considered her another mother.

Did she reside with you?

No, no—but close by, close enough that the children could walk it in a pinch. It wasn't easy walking distance, but it was walkable distance. When you have children and you have an active career, you have to make some compromises. My children grew up more independently than some. They were all very talented in music, so they were always going to music lessons. Most "music mothers" are with their kids, and they take them here and they take them there, and they watch them practice, and I didn't do any of that. The children went to their own lessons; they practiced themselves. The only thing we had was "family music." On Friday night, we would play together. That was my encouragement. When they were very little, I helped them with their lessons, because getting started in music is hard. But their schoolwork they did by themselves. I didn't monitor whether they did their homework or didn't do their homework. Yes, I got some flak from the teachers—that I was a negligent mother. But my kids went to the best schools. Two of them got Ph.D.'s—one in physics and one in math, not too shabby—and the others are doing okay. I can't complain. I don't think you have to hover over your kids so much.

Do you have a sense about how it worked?

I talk to them about it. They think it worked well.

. . . or why it worked? Maybe I should phrase it that way.

First of all, the kids were all the same age practically. I had four in less than five years. So they had a lot in common. They were close enough in age that they could do things together. So they could all go to the same camp; they were all in swimming lessons. The older one maybe found it a little boring, and the younger one maybe a little challenging. But they were there, because that was all the time I had. They coped with it fine. It was a little bit of a compromise now and then, but they made out all right. My daughter does it the other way; she gives the kids a lot of service. But maybe her kids need it. You know, every child is different. I can't say that she's wrong and I was right. But I did it that way because I didn't have much choice, and it worked out, more or less.[34]

As suspected, the theme that comes prominently to light is organization. For this woman, organization was aided by outside help, by a nanny who became a part of the family during over a quarter-century of care-giving. In many respects, this is an exceptional arrangement: such assistance is difficult to rely on over an extended period, indeed over a substantial portion of a career. This arrangement also is exceptional because it is costly and thus a privilege for many people.

As we would also suspect, ambition is evident. This is a woman who identifies strongly with and enjoys her work. She mentions compromises, though the exact nature of the compromises is not clear. There is a sense of compromise in the parental role (a "negligent" mother) and in the work role ("not very active on the outside"). But even in the face of compromise, the narrative is organized so that, in the end, things are "okay" and even very good by most comparisons.

Women are not unique in speaking of compromises. Men speak of them too, and what they say suggests how troublesome ambition sometimes is for work and family.

> When I first came here and in the first ten years or so, I think I could have done a better job in my personal life . . . my professional life was not balanced by some sort of personal life. . . . I think my wife was unhappy, and I think that affected our relationship. Then later, when I look back, it's something I regret in my life. Although our relationship is much improved and is much better—and it's very good now— I wish I had seen what I was doing or what I was not doing and had been better able to better balance my personal and professional life. So it's a regret I have now as something that happened in the past. Looking back, I always see that perhaps I should have balanced things a little better. . . . When at home my work always stayed with me in my head, and it still does to some extent. That's perhaps an aspect of being driven.[35]

Returning to the earlier quoted woman scientist, we also hear of a time when women were especially foreign in science. She received her Ph.D. in the 1950s, and her account of early experiences is unique among the women physicists. Blatant discrimination was par for the course.

Did you have a sense of what you wanted to attain [when starting out in graduate school]?

> No, I didn't. When I was a graduate student, I was told all the time, and pretty much by everybody, that there was no career for me. I had better treatment than most women, I would say, but I think that my mentors had no hope at all that I would have any career, and I didn't expect to have a career.

How was that conveyed to you?

> Very openly in many ways—in fact to the extent that I would ignore or avoid discussions with some of my advisors because, after a while, I didn't want to hear it. I was a good student when I was at———, and I was the recipient of a number of national fellowships. My mentor was

very upset every time I got one of these. If I was a guy, he would have been happy. But he had a different take on it—that being a woman, it was a loss of funds and loss of opportunity to somebody who would actually make use of the money.

Do you remember particular conversations or discussions?

Yes, they were very, very negative, and it had nothing to do with me personally, and there was no malice on the part of my advisor. In fact, about fifteen years after I got my degree he was totally won over, and he became an advocate for women in science. It was just a total flip. I was already a full professor here for several years when that phase-transition took place.

So how did this impact your notion of what you could do?

I assumed that I had no opportunity, and I was glad for every opportunity that I had, because I was anxious to work and continue my career. As a graduate student, I did quite well. I was pretty happy. I thought my thesis was pretty good, and I got quite a bit of recognition for it. Then I had an NSF postdoc after that.

How did you manage to get fellowships, awards, and recognition in graduate school and thereafter when you had so little support?

I didn't have little support; I didn't need much support. I had all the support I needed.

Support from your advisors?

[Sharply:] You don't need support from your advisors to work.[36]

The three remaining accounts by women scientists differ from this one. In what some might view as surprising, little mention is made of discrimination against these women or against other women they have known in science. Whereas discrimination is in full vigor in the account above, it assumes a comparatively minor role in the account below. This is the account of the younger elite, who was the only other women to raise the issue of discrimination in any way. She is thirty-six.

You are one of the few women on physics faculties. How do you account for being where you are today?

I've thought about that a lot. I'm seeing it from the other side too now. I'm working with female and male students which is very interesting. I think there are differences in the way people are raised. The aggression factor is in there somewhere. And I think there is prejudice. There has been a long history of women not having the same positions as men.

Were you encouraged to apply to a graduate program in physics?

Yes.

So you had support?

I had a lot of support. That made a huge difference. And practically too: family helping with children.

Did you have kids while you were on the faculty here?

My first one was born a year before I came here, and the second one was born nine months later.

Can you think of any major junctures where things might have turned out differently as a result of being a woman?

That's a good question. I must say, I don't know if this is naive, but since I've come [to this university], and even before, I have rarely felt that I'm being treated differently because I'm a woman. I'm actually a little surprised. Colleagues don't seem to talk about it.

What about at the graduate school level? Since you are now in a position to see, do you think the socialization process between male and female graduate students is different?

That's a good question. Part of it is already set up. Women usually come in with less lab experience. I try to fight it; I try to encourage the women to do things. So, yes, it still goes on. And also there are other things, subtle or not so subtle. When I was a graduate student, all the male graduate students in the building came to my office on a field trip to see the woman graduate student. That makes you feel a little strange. That was ten years ago. Maybe it has changed. But I bet it hasn't completely.[37]

Despite presenting discrimination in low key, this woman nevertheless raises the issue. Her comments are consistent with patterns explicated by the most recent large-scale research on gender and science: discrimination seems more subtle now and, though far from being completely eradicated, has begun to wane substantially (Sonnert and Holton 1995).[38]

Like the first woman physicist quoted, this scientist emphasizes support for children: extended family play a role in maintaining a work-family equilibrium. On ambition, she had simply this to say: "It's been important. It's not easy sometimes—having two children and a job like this." As suggested, one suspects that without ambition or organization, a woman would be much less likely to achieve success on all scores.[39]

The role of ambition comes across clearly in the account of the communitarian, the third of the four women. Thirty-nine years old, she is married but does not have children. Nowhere in the interview did she explicitly mention her role as a woman scientist. Without prior knowledge, a reader would have difficulty ascertaining the gender of the person quoted. The account below is relevant, however, because it tells how this woman, much like the others, views her career.

Would you say that you're ambitious?

Yes, I think I am. Again, I'd be lying if I said I'm not.

Do you feel it's bad to admit that you're ambitious?

Well, I suppose being ambitious can be interpreted by some people as being "too much" [as in overeager]. And so maybe I'm too ambitious.

What is your ambition?

It's to keep climbing all the ladders and to go up to the highest one.

Where do you think that ambition comes from?

I guess I've always been that way. It's a part of me. I've never been happy with: "I could stop now." I know that I'll be tenured next year, so I could go on and have fun. But it's not enough for me. It's not so much to keep a job; I just need to continue to go up. My father has always been demanding. My brothers and sisters all have fine professions, and so he's pleased with that. And he stressed that. I was raised by a father who made clear what his expectations were, and [he] continuously reminded us. So it might be that originally I wanted to please him. I distinctively remember being competitive and wanting to be the first in my class.

What role do you think ambition has played in your life?

I think it's a big determinant. I'm quite competitive, and it's by being competitive that I've gotten to where I am. It's not easy. It doesn't come easy; you've got to go to it.[40]

The fourth and final woman—the pluralist—raises her gender in the remotest of possible ways. Forty-five years old, she is not married and never has been and does not have children.

Has there been some inner conflict or personal turmoil that you have sought to understand in your life?

Maybe it's a good side or a bad side of being so busy with research that it just takes too much time to ponder those things. I don't have a family. But people who do, I'm sure they have many more—when they set their priorities—[I'm sure they have more] time constraints. I'm sure it sometimes must be hard.

Has not having a family been a source of concern?

I worried that I should be worried. . . . I worry that's maybe something I should be worried about. It's not anything I notice. I enjoy friends and things like that. . . . I'm so used to being alone that now it has come to the point that having at least a little bit of privacy is very important to me.

Did you at some point make a conscious decision that this was the kind of life you were going to have?

No, I don't think so. It just happened that way. As I say, I feel fortunate that I have a job, I can support myself, and it's also possible now for a single career woman to function and also to do anything she wants.[41]

The account is telling. This woman is not worried about being single and childless (cf. Levinson 1996); if anything, she is worried about her lack of worry. And like the women whose accounts precede hers, she shows ambition and commitment: she enjoys her work.

Men, Women, Career, and Ambition

Historically, ambition has been a masculine concept. Consider the ways in which the word "ambition" is used. It is a long-standing referent to occupation; it does not customarily apply to family roles. One is an ambitious attorney, physician, or athlete but not an ambitious father or mother, wife or husband. By contrast, as a father, mother, wife, or husband, one is ideally loving, supportive, and cooperative. Because female participation in the labor force has increased significantly only since the 1960s, ambition in historical perspective has most readily applied to men and their work (cf. Hughes 1958).[42]

In examining how ambition figures into women's career narratives, I suggested it plays just as prominent a role as in men's. What differs is the imperative for organization. Is this a satisfactory conclusion? The question remains as to precisely how we understand ambition in the context of gender and work.

Additional comparison between women and men sheds further light. Specifically, let us compare several objective characteristics of role performance. In making comparisons, I again alert readers that caution must be exercised in overgeneralizing on the basis of a subsample of only four. Here, as above, comparisons are suggestive, not definitive.

In previous chapters I presented information comparing key aspects of scientists' role performance, including how their performance differs across cohorts and across academic worlds. To place these women in greater context, we should compare their individual performance characteristics with the group norms. Comparison on several measures establishes some solid, though still slightly shaky, ground on which to gain greater leverage on the issues at hand.

We begin by examining the characteristics of the women's careers when they were starting out in science. In table 14, I compare the women's publication productivity with group norms at the time of taking the first full-time academic job. The table shows a clear and consistent pattern. All of the women entered the academy as unusually productive

Table 14

Productivity Comparisons between Women and Corresponding Cohort
at Time of Academic Employment

Woman	Number of Papers[a]	Cohort Mean
Elder elite	18	5.2 (Cohort I)
Younger elite	14	12.5 (Cohort III)
Pluralist	12	7.2 (Cohort II)
Communitarian	10	12.5 (Cohort III)

Note: See table 11 for equivalent information on the whole sample. As previously used, "time of academic employment" means the point at which the scientist first became an assistant professor or visiting assistant professor.

[a]Reflecting the need for caution, the numbers of papers for the women should be interpreted in context. The cohort norms reflect the productivity of individuals entering science in different years. That is, in any one cohort, individuals received their Ph.D.'s over a span of several years, resulting in variant levels of productivity.

scientists, having published a significant number of papers in both the graduate and the postdoctoral phases of their careers. At the time they took their first academic posts, they had outproduced the average in their respective cohorts. This is all but certain for the communitarian, despite the fact that table 14 lists her as having 10 published papers, compared with her cohort mean of 12.5. The cohort averages do not separate contextual effects: the averages include elites, pluralists, and communitarians. If we compared this women only with other communitarians, and then only with those in the same cohort, she would outproduce the average.

Differences from the cohort are especially remarkable in the elder elite. When she took her post, she had published 18 papers, compared with a cohort mean of 5.2. This is noteworthy because she entered science when bias against women was still strong; indeed, very few women ventured into science whatsoever (Rossiter 1995; Zuckerman and Cole 1975). From her own account, we can explain her superior performance by an unusual degree of institutional support, at least relative to other women at that time. In addition, she is particularly ambitious. Finally, the fact that she held a research post for seven years between her postdoc and her academic position also accounts for her relatively high productivity.

The productivity of the other three women scientists was likewise enhanced by their postdoctoral appointments. But then, we must draw the same conclusion for the men, the majority of whom also held postdoctoral positions (especially in cohorts II and III; see table 11). Consequently, although we are dealing with only four people, there is substantial evidence that these women entered science with especially strong publication records.

In a similar vein, we may compare each of the women with the larger

Table 15
Total Productivity and Quality Comparisons between Women and Sample

Woman	Total Number of Papers	Cohort/World Mean[a]	% in Major Journal
Elder elite	272	109.3	60.7
Younger elite	24	29.3	95.8
Pluralist	44	73.3	100.0
Communitarian	26	21.7	65.4

[a]The mean is of cohort and world. Thus, the elder elite corresponds to Cohort I, Elite World; the younger elite corresponds to Cohort III, Elite World; the pluralist corresponds to Cohort II, Pluralist World; and the communitarian corresponds to Cohort III, Communitarian World. See table 3 for equivalent information on the whole sample.

sample on other performance measures. Table 15 compares the total number of papers the women have produced to date with the average for their respective cohorts and worlds (this is a quantity comparison). In addition, table 15 gives the percentage of the women's papers appearing in major physics journals.

The elder elite, highly productive in her early career, has been remarkably productive throughout, publishing a total of 272 papers compared with 109.3, the average for other elites in her cohort. This woman's total number of published works not only vastly exceeds the average but also establishes the upper limit of all published works in the entire sample (see table 3). She is our most productive scientist. Considering the historical place of women in science, there is some irony that in a study of scientists, our most productive one is a woman. Equally significant, a majority of her published work has appeared in major journals (60.7 percent).

A similar pattern is evident for the communitarian. She has been more productive than her communitarian counterparts within her own age group (publishing 26 papers against the mean of 21.7). In addition, a majority of her work has appeared in major journals—65.4 percent.

The younger elite and the pluralist exhibit slightly different patterns. Each has produced less than the average for her cohort and world. The elite has published 24 papers, whereas the average among elites in her age group is 29.3. Similarly, the pluralist has published 44 papers, whereas the average among pluralists in her age group is 73.3, a substantial difference. But notice the quality of publications for both women. Although they have produced less than the average, the vast majority of their work has gone to the best journals. The younger elite has published 95.8 percent of her work in the most competitive journals, and the pluralist has published 100 percent—all of her work—in major journals.

This pattern—publishing work less than or near the average in quantity but higher than the average in quality—is consistent with the latest research explaining men's and women's productivity differences by "scientific styles" (Sonnert and Holton 1995). Evidence suggests that women take a more cautious and perfectionist approach to their work because of their marginal status within science. Few in number, and thus without major systems of support or authority, women more strenuously uphold traditional standards of science, such as carefulness. They are more likely to conform to extra-high standards in order to be viewed as serious and legitimate members of a community. Making "mistakes"—or simply not making a majority of significant contributions—may have higher costs for women than for men. The costs of their perfectionism may be seen in their productivity.

Ambition Revisited

Looking at role performance, we see that the women scientists are, on balance, highly successful. This holds even at the earliest points of their careers. Their narratives highlight ambition. The women also often note the necessity of "being organized," and sometimes of seeking compromise, in how they conduct themselves in and outside of science.

The findings suggest that women are held to higher standards. It is possible that only the very best women enter the professorial ranks. Other reports partially support these claims. Studying all scientists, researchers have consistently found that a higher proportion of women than men are involuntarily unemployed (National Science Foundation 1990; Vetter 1987).

Perhaps one of the most interesting conclusions to draw is the high degree to which these women resemble men—both in how they perform and in how they narrate their careers. Here, we reach a more qualified conclusion. In performance and in narrative, the range of variation among the women is slim, narrower than the variation range among men. If not elites in the institutional sense, all of the women are *elite-like*.[43] They are elite in how they perform and elite in how they sound. In saying, therefore, that these women resemble men, we may more specifically assert that they resemble a subgroup of men: elites. This in turn leads us to conclusions about ambition. In all probability, these women are more ambitious than *most* men, but they are not necessarily more ambitious than the most ambitious men. It is the most ambitious men whom the women most closely resemble.

The bulk of this chapter has been concerned with turning ambition "on its head"—with seeing ways in which people are driven by self-doubt as opposed to dreams. Looking at these women's self-doubts, we are again led to signs of their overarching "eliteness." Nagging them are the same concerns that most typically nag elites: concerns about time, progress, and standing. For the women with kids, these concerns are joined in the same or next breath with anxieties about performing both work and parental roles effectively.

Each woman answered the question, "Have you had doubts about yourself?" The elder elite answered:

> I find it pretty hard to say no to people, so I tend to get asked to give too many talks, and too many this and that. I work a lot of these things on what I call a time-allocation basis. I have certain things that I want to do—like my research—and then I have things that I'm asked to do—like giving talks. I figure, okay, I'm going to spend so much time on this and that's it. . . . That's the way I've operated for the last fifteen years or so, when my schedule has gotten pretty much out of control. . . . I could turn around and say I'm only going to do a third of things and do them well. I'm not sure that's the right answer either. I don't know what the right answer is. I haven't figured it out.[44]

She added:

> If you can put me on the spot on this, I could say that when my children were really young I would ask myself, "Should I be working?" All the input I got from everybody, [except] my husband, was "Why are you doing this, anyway?": family, teachers from school, neighbors. I kept going. I thought I was doing okay. You get so that you shut it out. I got to the point when reading child[care] books—the child has this and that symptom—you try to look it up and educate yourself about what's going on. I would only read the very specific thing that had to do with the ailment because the general philosophy—as soon as you read it, you feel like a criminal. The books that were written in the fifties about child-rearing have the mother staying at home with the kids, otherwise she is a traitor. I got so that I read only the pages of [Dr.] Spock that I needed to. My kids did better than Spock's kids. And I think my kids liked me better than Spock's kids liked him.[45]

The younger elite said:

> . . . family issues: "Am I causing my children to become psychopathic murderers or something because I'm here at the office rather than at home?" . . . What I find, and I think it's also true of other women in this position, is that you go on in the years and you actually see that your kids are doing okay. My son is six years old now, and he's well-

adjusted. So after a while I think that reinforces that what you are doing is not that awful.[46]

The pluralist answered:

> People are aware of how little they know. I have students or postdocs around because they ask questions all the time, and that forces you to admit how much or how little you know. You hear some physicist give talks or you see a paper written by a physicist, and there are times when you know there's no way you would have been able to do that same kind of work. In that sense, you ask, "What am I doing to compete?" [People] hope that at least they are contributing something—[that] something will last.[47]

Lastly, the communitarian said:

> There are quite a few things I would like to be better at, but I don't have any time—I cannot devote [time] to it. We bought software for our computers to run the experiments, to control the experiments. It takes time; you have to learn it—you have to sit down and learn it. I don't have time to do it because I have to write proposals, I have to keep up with papers. I have to do many other things, and there isn't enough time for me to tackle something like this. I don't know anything about it; I wouldn't know how to run it; but I would want to know everything about it. I'm realizing now more and more that I do not know everything, and that's frustrating.[48]

Whether evident in women or men, these concerns stem from the institutionalized drive for recognition in science. But it is the elites, and those who are elite-like, who most often mandate and adopt this drive unhesitatingly.

Conclusion

This chapter examined the inner voices that guide scientists' thoughts, feelings, and actions. I have focused on self-doubt in order to see how insecurity about status gets channeled into ambition and to determine which groups are most likely to experience this process. I have explained why we find compensatory achievement patterns in some groups more than in others. The answers, I argue, lie in the cultures of the worlds in which scientists work.

Psychosocial anxieties inform our understanding of the social worlds in which people interact. As manifestations of moral orders, psychosocial anxieties set worlds apart in systematically different ways. Self-doubts connected to work are most apparent among elites, who doubt progress, standing, and status. Such self-doubts are apparent but less pervasive among pluralists and are apparent only in anomalous communitarians. In strictly objective terms, the further "up" one goes in the institutional hierarchy, the more one finds individuals with the same anxieties. The most ambitious are also the most self-doubting, the most insecure, the most inadequate in their own minds, the most vulnerable. They are conflicted, uptight, impatient, anxious. Pluralists and especially communitarians are in greater harmony with themselves, explaining why professionally they are so more often at ease. Women scientists, both in how they narrate their careers and in how their careers look objectively, are most similar to the most ambitious men.

On the one hand, self-doubts drive individuals to the top ranks. Thus we find the largest concentration of self-doubters among elites. Even outside the elite world, self-doubters rise to the top, becoming elites in a pluralist or communitarian world. On the other hand, elites *induce* self-doubts because of their competitiveness over recognition. Regardless of career stage, elites validate their competence through competitive efforts. Progress becomes an imperative and an achievement—a vital symbol in which people stake their self-identity.

SIX

Voices and Visions

I have devoted the previous chapters to uncovering the role that ambition plays in careers. I laid the theoretic ground for the study in the introduction and explained my sources of data and my research methods in chapter 1. In chapter 2 I described the three academic worlds and several ways in which they differ. Each of the subsequent chapters addressed how ambition manifests itself in people in the different worlds and how ambition changes over the life course. That is, I located ambition in the *times* and *places* through which people move. In doing so, I addressed how individuals create and re-create self-identities, how they know and understand themselves as their lives unfold.

Chapter 3 discussed the ways in which scientists account for their pasts. The scientists described their passages from graduate school to their early careers and ultimately to where we find them now. From these vantage points, we saw ambition's chronology, the forms and intensities it took over time.

Chapter 4 presented and examined the outlooks that the scientists have formed of their futures. We entered their worlds of dreams in order to grasp their ambitions and to further delineate how these ambitions differ by time and place.

In chapter 5 I turned ambition on its head. Before chapter 5, we viewed ambition through positive self-attributes, particularly those that compose individuals' scenarios of success and grandeur as high-aspiring scientists. In chapter 5 I bridged these views with the scientists' accounts of their vulnerabilities about status. I did this on the premise that insecurity rounds out the picture of where scientists draw their ambition. Self-doubt often plays a demonic-like role, fueling ambition on the one hand while threatening and at times killing it on the other.

In this chapter, I further locate the scientists' ambition in its sociocultural context. I will elaborate the social and cultural conditions that inculcate ambition, specifically what I have called "ambition for greatness." The academic world is not the first place that instills ambition in a person's life, of course. A successful passage through preprofessional rites presumes a drive in addition to ability. Most, if not all, scientists have entered academic worlds with ambition of some kind, though the types of ambition vary, as we have seen. The larger point concerns what time and place do to ambition—and consequently to people's lives.

From the outset, my approach has been highly inductive. I have relied heavily on the accounts of individuals to address the ways in which their worlds of work differ systematically on a number of dimensions. Individuals therefore have provided entré into both the self and the environments in which they are socially situated.

Environments, I have argued, radiate cues about conduct, prescribe a course of action, and outline an overall way of knowing a world and one's place in it. People's beliefs and actions inform our understanding of the social and cultural conditions that undergird their worlds. Although I will introduce select passages from scientists' accounts in what follows, this chapter is largely the product of the reasoning made throughout the previous chapters.

I will divide my discussion into four parts. First, I will discuss the structural and cultural characteristics that differentiate academic worlds. We may take these characteristics to more generally distinguish worlds of the academic profession writ large. Second, I will explain how these structural and cultural differences yield different narratives, and I will introduce the concept of the master narrative. Third, I will discuss these narrative distinctions in light of how we may view science as a profession—as an array of narratives. I will conclude by addressing how sociocultural variations shape ambition in different ways, leaving one form—greatness—as a pad from which careers are launched and lived out.

The Internal Differentiation of Science

Beginning with chapter 2, where I discussed typical careers, and proceeding through chapters 3, 4, and 5, where I discussed how ambition manifests and fails to manifest itself in individuals, I have delineated the constitutive characteristics of the three academic worlds—communitarian, pluralist, and elite—and how they are set apart from one another. At the same time, however, I have addressed the ways in which the boundaries of these worlds sometimes cross over one another.

I have argued that the dimensions along which worlds differ are best considered in both structural and cultural terms. These structural and cultural distinctions represent ways to differentiate not only this sample of worlds but the population of elite, pluralist, and communitarian worlds that make up the whole academic system.

Structurally, worlds differ first in the number and range of *opportunities* available to people. Ambition varies across worlds because different worlds offer different opportunities for ambition. Worlds vary in their wealth of human, fiscal, and infrastructural resources. These differences produce inequities (both in number and in quality) in opportunities to get ahead.

Second, worlds differ structurally in the predominant systems of reward or *incentives* extended to individuals. Definitions of achievement vary across worlds, so that in one (especially the communitarian) world, achievement in teaching or in administration carries equal weight with achievement in research. Failings in any one of these roles, including research, is excusable, though failings in *all three* normally results in sanction. By contrast, in another world (especially the elite), achievement is normally defined by research. Elites most easily forgive failings in administration and/or teaching. Failure in research, however, is like a knife in the side.

Culturally, worlds differ first in the *expectations* governing role performance. People hold different beliefs about what constitutes "achievement." Worlds have their local definition. Elites mandate greatness. Pluralists and communitarians, by contrast, gravitate toward other and more pragmatic outcomes; greatness becomes an increasingly all-too-distant dream.

Second, worlds differ culturally in the *symbols* that characterizes careers said to be standard. Elites instill in one another ambition for greatness if only because greatness embodies the pantheon—the heroes whose photographs line the corridor walls, who come to the meeting rooms to give talks, who have trained great minds that in turn have trained the current elites, and who are, in effect, worshiped for their superior scientific powers. Greatness becomes the standard, the baseline from which all elites (or at least most) are striving. In other worlds, greatness is a more far-removed curiosity. Scientists speak of heroes not as colleagues, collaborators, or associates—let alone as friends—but as distant, nearly deified characters who arouse general interest. Interviews with communitarians revealed symbolic faux pas: names of scientific legends were mispronounced; other legends were forgotten or not known. Such "mistakes" highlight the sociocultural distance between elites and communitarians. Moreover, they lend still greater weight to my metaphorical characterization of the two groups: elites and communitarians live *worlds apart*. All scientists may share the language of mathematics, but in keys ways their dialects reflect the different worlds of the profession they inhabit.

Boundaries between worlds arise along each of these four dimensions. We may therefore speak of different worlds of opportunities, of incentives, of expectations, and of symbols. And we may just as easily speak of opportunity boundaries (limitless or limited), incentive boundaries (many or few, narrowly or widely defined), expectation boundaries (high or low), and symbolic boundaries (grand or unornamented) (cf. Lamont 1992).

When we speak of such worlds or of such boundaries, however, we must also be aware that their attributes are highly intercorrelated. One attribute often gives rise to or feeds off of another. Where we find a world of plentiful opportunities, we are also likely to find a world of potent incentives, of high expectations, and of mighty symbols. Where we find a world of meager opportunities, we are also likely to find a world of faint incentives, of low expectations, and of mild symbols. One condition predicts the others. At root, these structural and cultural distinctions account for and describe the internal differentiation of a profession.

Introducing the Concept of the Master Narrative

This study has shown that structural and cultural differences engender distinct ways of accounting for professional passages. Elites, pluralists, and communitarians use different rhetoric to publicly present their life courses. Their narratives are shaped by sociocultural forces at play.

To conceptualize these distinctions, and to thereby achieve a more abstract and generalizable understanding, I am setting forth the idea of *master narratives*. Master narratives refer to the dominant communication patterns that characterize how people order and present their passages through time and place (cf. Buchmann 1989; Gergen and Gergen 1984; White 1993).[1] Accordingly, they exist as a prime cultural tool with which to understand, and gain leverage on, the dynamics of identity in all social life (cf. Bruner 1990; Ricoeur 1983–85; White 1987).

Narratives of Being, Narratives of Becoming, Narratives In Between

We have heard people from the same social world narrate their careers in similar fashions. Regularities in their rhetoric flow from regularities in their environments: regularities in opportunities, incentives, expectations, and symbols. Carving careers from what is on hand, people tell narratives of being or becoming.

Elites see themselves and are seen by others as pushing to get ahead, their self-identities competitively staked in attainment and recognition. The life course consists (in principle and normally in practice) of a steady upward climb. They would appear to possess the Olympic virtues: *citius, altius, fortius.*

Communitarians describe themselves in ways that highlight steadiness. They often "get stuck," largely because the four sociocultural conditions in their world do not favor upward mobility. After early career phases, little social control regulates or helps to inspire peak performance. Their master narrative underscores this leveling-off.

Pluralists portray themselves in both ways. Some embrace the idea of becoming, as most typical of elites, while others embody the principle of being, as most common among communitarians. Still others are hybrids of the two forms. Their master narrative stresses this mediation. The most typical pluralist strives in early and midcareer and levels off thereafter. Calls for greatness in any one phase are not usually answered, nor are they even always heard. This is particularly evident in later stages, when pluralists decelerate as elites continue their upward climb.

Just as sociocultural conditions explain differences between worlds in how individuals account for their journeys, they also explain similarities. Throughout these chapters I have discussed the ways in which academic worlds differ internally. I have explained how individuals who break away from the norms of one world begin to resemble some people in another. Highly successful pluralists often resemble certain elites; more modest pluralists often resemble certain communitarians. In short, there is overlap.

There is overlap because at a general level, the academic worlds are oriented toward similar purposes. At a general level, their members are united by common activities. This is reasonable to expect: all three worlds are universities. But though there is similarity at a general level, we have seen how individuals embrace these purposes and engage in these activities in varying degrees and intensities. Prevailing structural and cultural conditions are therefore *more or less* applicable to any given world; they are not strictly delimited.

Thus although master narratives correspond to each of the worlds, individual narratives sometimes depart from the master form. People are not, of course, perfectly uniform in how they account for their passages, largely because they experience their careers in different ways. People accelerate and decelerate in their careers at different intervals; they encounter turning points sooner or later than others, and some experience certain turning points that others fail to experience at all.

These differences surface as a result of structural and cultural overlap. For example, a subset of pluralists may be exposed to many of the same conditions common to elites, and so their careers look and their narratives sound like those of elites. At the same time, some of their pluralist colleagues fail to be exposed to these conditions. This failure may stem from performing weakly in the past, or it may stem from being at an advanced career stage in which people subscribe to a set of expectations that are outmoded. As a result, their careers and narratives resemble those of communitarians. The same processes are found among elites and communitarians: some more closely approximate people in other worlds.

Consequently, in addition to the master forms of accounting for typical journeys are accounts that represent variations on the locally dominant theme. These may best be referred to as *branch narratives:* accounts that follow the master narrative (the trunk) up until a given point before breaking away, at which time they begin to resemble the accounts of those who inhabit an adjacent world.

Similarity and Difference the World Over

It is perhaps useful to underscore these points through analogy. The problem is this: how can any phenomena be described as different in fundamental ways yet also similar? Moreover, how can a conceptual scheme—like the one I have adopted here—hold up when it seeks to explain difference and at the same time account for similarity? Isn't such a scheme self-contradictory?

The answer is no. We can understand the distinctions among academic worlds in terms of dominant patterns, what statistically would be characterized as central, gravitating tendencies.[2] At the same time, however, there is no theoretically sound reason to expect complete uniformity in any one world. If that were what we found, we would indeed be dealing with a sociological marvel: a world of no variation. Such would, of course, be highly suspect.

The worlds differ internally because the people in them differ. Boundaries exist between worlds, but they are not impermeable. Nor should we expect academic worlds to be completely distinct (except perhaps at the polarities) because, as explained above, at a general level they are similar: all are universities, and by virtue of that fact alone, they are bound to have characteristics in common.

I may be dealing with apples, oranges, and pears, but in each case I am looking at fruit. In many fundamental ways these fruits differ (in color, size, taste, texture). But they also share characteristics (they all have seeds), as one would reasonably expect among members of a single family.[3]

Used to characterize regions of professions, the "world" metaphor itself helps to explain how we have come to look at similarity and difference in our subject. Take several parts of the world—several nations—and describe life within them, as a good many people have tried to do. One nation may be described as pluralist, another as communitarian, and another as something else. These are central tendencies, overarching ways in which to describe behavioral patterns and social conventions applicable to a wide segment of each respective population. But though we may use such categories to describe and explain a significant portion of social life flourishing within those groups, there is variety in each one. What is more, the range (or distribution) of variety may itself vary from world to world (elites are more homogeneous than pluralists). Some of this variety may not conform to what is described as the locally dominant set of patterns; instead, it may more closely approximate patterns of other worlds.

Seymour Martin Lipset, among others, has engaged in just this kind of exercise (Lipset 1996). He has sought to describe and account for key characteristics of U.S. society, which, incidentally, could well be described as pluralist. Lipset finds that there are dominant characteristics that distinguish Americans while at the same time he observes subpatterns that speak of internal differentiation. In some cases, subgroups more closely resemble other societies—those from which they originated or to which they are in other respects closely allied (such as blacks' identification with African regions or Hispanics' ties to Latin America). Nevertheless, these groups are bound to an overriding collective, national identity: a set of standards and practices that come to define what it is to be "American" (cf. Anderson 1983).

Accounts of internal variety do not undercut central characterizations—elite, pluralist, communitarian, agrarian, industrial, postindustrial, or something different still. Indeed, central characterizations are the result of the observation and consideration of respective internal conditions (hence American society gets defined as pluralist, whereas Japanese society gets defined as comparatively homogeneous).[4] The more scientifically responsible route accounts for variation subsumed under a general heading rather than proceeds with a characterization of dominant patterns that ignores deviations from master themes. Variation fills in the picture. It is precisely this concern with central currents enjoined with countervailing ones that has guided my conceptual view.

Sources of Overlap

Prior discussion has suggested sources of the overlap among academic worlds. First, academic worlds are fundamentally alike. We are not comparing orchestras, car dealerships, and laundromats. Analogously, we are comparing types—and very different ones—within a respective system of orchestras, car dealers, or laundromats. Accordingly, the various types are linked within a system by common purposes, although those purposes may remain at a general level as each world fulfills a role in a specific niche. Hence we have "research" and "comprehensive" universities and "liberal arts" colleges—all espousing different missions and exhibiting contrasting characteristics (see Carnegie Commission on Higher Education 1994). But at the same time such schools operate on a common ground of transmitting and extending knowledge, forming a system of education.

The overlap, however, has deeper roots. As I have previously suggested, overlap among academic worlds, to the extent it occurs, is due to two primary factors: market conditions and doctoral origins. The first operates as a force at the macro level, the second at the micro level. The first functions in response to supply and demand, the second as a corollary of socialization. Though analytically distinct and sometimes operating apart from each other, the two factors also at times work in tandem, as I will explain shortly.

The first factor, market conditions, causes overlap among academic worlds when market forces propel similar people to different types of places. Historically when jobs are highly scarce, institutions of all types are able to benefit from a flow of highly research-oriented candidates. This encompasses the process of "institutional upgrading," discussed in considerable detail in chapter 4.[5]

Over the past several decades, as market conditions have tightened, pluralist and communitarian institutions have become more research-oriented. This trend stems from their ability to hire prime candidates in a highly competitive labor pool. In the past—when market conditions were more fluid—candidates, especially candidates with elite doctoral origins, had greater discretion about the places where they chose to work. The trend also reflects an awareness of where prestige most prominently lies in the education system—with research, development, and technology—and reflects attempts by institutions to gain more of this prestige, becoming more publicly visible and more financially prosperous.

Though market conditions and the attending effects of institutional upgrading partly account for overlap, all indications suggest that overlap has had a staggered character. That is, pluralist and communitarian institutions have not engaged in upgrading practices in synchrony with one another. When market conditions first tightened, pluralist institutions were the first to benefit. Compared with communitarians, pluralists experienced working conditions more closely approximate to those of elites; thus, when push comes to shove, those with elite doctoral origins are more likely to see a home among pluralists. Consequently, people with elite doctoral degrees are more common among pluralists than among communitarians.

As market conditions tightened further, communitarian institutions began to attract people with elite doctoral backgrounds. Thus, when one looks at individuals who deviate from pluralist and communitarian norms, there are clear patterns: "elite pluralists" tend to be at early or midcareer; "elite communitarians" most often are young, in the earliest phases of their careers.

Institutional upgrading must be understood in the context of school missions (Meyer 1970). To a degree, specific missions may change as schools modify their activities. But this is a delicate situation. Any institution that seeks change realizes that change cannot come at a significant expense to enrollment. Naturally, diminished or chronically unstable enrollment pools foretell institutional demise.

Consequently, institutions may "upgrade," but their missions remain faithful to teaching. Pluralist and communitarian institutions have become more research-oriented, embracing the research model of the university with greater vigor, but this does not mean they are identical to the elite research universities. Their commitments to teaching remain strong because they usually must rely on local or regional applicant pools to survive. Overlap has occurred, then, but within institutional limits. For these reasons, we would never expect to see uniformity throughout the education system, and any overlap among academic worlds always exists at a threshold: in fundamental ways, academic worlds serve different clienteles.

Market conditions interact with the second factor yielding overlap: the doctoral origins of individuals. A highly contracted labor market spreads candidates with elite doctoral backgrounds to more and more varied types of institutions. Yet even when market conditions are relatively flexible, overlap among academic worlds still occurs. This leads us to look further at doctoral origins, which sometimes operate as a force independent of market conditions. As Harriet Zuckerman has observed, conceptions of the scientific role, styles of work, and standards of performance develop during graduate training; thus school background may help to explain differences in self-identity and orientation to work (Zuckerman 1977).

As part of their graduate school experiences, individuals are socialized to hold particular career expectations. For example, scientists with elite backgrounds are normally socialized to aspire to greatness. Elites establish greatness as a possible career outcome but above all as one toward which all serious newcomers should strive. This is true in pluralist and communitarian training grounds also, but in lesser degrees, in part because the expectation is less enforceable: greatness is not a norm; nor is it something that is easily crafted from the resources or people on hand.

Overlap among worlds potentially results when scientists holding elite doctoral degrees make their way into pluralist and communitarian institutions. Some scientists are able to conduct careers in ways that approximate the careers of elites, and they narrate their careers in a similar manner. They may be able to partially withstand the otherwise overriding contextual effects by having a small yet critical mass of similar individuals in their immediate work environment. In addition, they also are likely to have connections to elites through collaboration that extends outside of their local institution.

Doctoral origins, however, are more complex. Although pluralists and communitarians with elite doctoral origins sometimes embrace an elite-like career, others with similar origins do not. Furthermore, a handful of pluralists and communitarians *without* elite doctoral origins embrace an elite-like career and tell an elite-like narrative. Even though doctoral origins account for a high degree of the variation, they do not account for it all.

Other factors, often connected to doctoral origins, also play a role in producing overlap. These factors can be grouped into four interrelated categories: mentor characteristics; extent of research collaboration; productivity; and personal experiences. The first three of the four factors are the most closely intertwined. It is conceivable that the most ambitious pluralists and communitarians are those who had graduate school and/or postdoctoral mentors who were themselves ambitious (Fox 1983; Reskin 1979). Thus, whether a scientist has elite, pluralist, or communitarian doctoral origins, individual achievement expectations get defined by the status and identity characteristics of mentors. This would explain why some pluralists and communitarians exhibit ambition for greatness and others do not. It would also explain why a fraction of those without elite doctoral origins nevertheless exhibit elite-like qualities.[6]

The effects of mentor-student relationships can be particularly instrumental when the ambitions of mentors are substantially greater than those of the overall collectivity, such as highly ambitious mentors in pluralist and communitarian settings. The mentor not only is in a position to inculcate similar attributes in professional offspring but also is likely to have key outside connections that facilitate mobility and elite identity-formation in students.

This brings us to the second category. Mentors can facilitate collaboration, not only with the mentors themselves but also with others throughout subsequent phases of an offspring's career (Reskin 1979). In effect, mentors help launch students on a trajectory in which access to opportunities, elite expectations, rewards, and symbols is readily at hand. Budding scientists thus come to work in a microworld that in many ways affirms an elite identity, even though they may call the pluralist or communitarian world their home (cf. Crane 1965; Long and McGinnis 1981).

In turn, collaboration fosters research productivity, the third category of factors producing overlap. Productivity itself creates a further ground on which to make claims to an elite self-identity, even if an individual is not based in an elite world (Pelz and Andrews 1966). Thus, mentor characteristics, collaboration, and productivity are closely linked to one another. It should be observed, however, that in principle, collaboration and research productivity also operate independent of "mentor effects" (Fox 1983); some people form collaborative ties and are productive without having had a mentor play an instrumental role in their careers.

Finally, personal experiences account for a portion of the overlap. This was especially evident among elites (though not confined to them). The elite world is defined as one in which every aspect of the professional stage has been set for superior achievement, and indeed all elites, in their own ways, are or have been superior achievers. Yet we heard the accounts of elites who decelerated sharply; they drew away from work. Such individuals assumed many of the characteristics more common among communitarians and some pluralists: withdrawing from research and emphasizing teaching, including assigning a value to the interpersonal dimensions of the teacher-student relationship. In all of these cases, personal experiences—divorce, death, ill health, family concerns, psychosocial developmental issues, and even early success that caused the career to "peak" too soon[7]—led people down alternate paths. Instead of embracing the moral career, people reset priorities. In doing so, they often found themselves exploring other aspects of the human terrain, usually those that involved a renewed appreciation of relationships and contact with others.

The Ecology of Careers in Science

Theoretic Perspective

The social order of the science profession is best conceived as a continuum of three overlapping spheres, each sphere representing one of the academic worlds. The Venn diagram in figure 1 represents a *theoretic* depiction of how science is ordered. On this theoretic level, the worlds are spaced evenly apart. Based on empirical findings, this arrangement reflects the idea that the worlds penetrate each other's boundaries. Structural and cultural conditions characterize each world, but to a degree they bleed over into another.

It is important to note that the extent to which worlds overlap is an empirical question. We can determine the extent of overlap by mapping the characteristics of any given set of universities, which this study has done for a set of six. For example, a given pair of universities (Stanford University, an elite, and Bowling Green State University, a communitarian) may not overlap at all. Others (Cal Tech, an elite, and the University of Massachusetts, a pluralist) may have considerable overlap—at least the extent of their overlap may be greater than that between Cal Tech and some other pluralist institution, but as usual this remains an empirical matter.

The diagram shows that pluralists, as I described throughout this book, constitute a hybrid of elites and communitarians. In this world we find individuals whose careers resemble those of elites or communitarians or something in between. At either end of the spectrum are polar

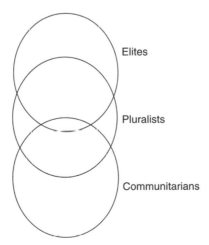

Elites

Pluralists

Communitarians

Figure 1   The Social Order of Science Careers (Theoretic Formulation)

types of people. On the elite end are the individuals who are the most devoted to (and the most successful at) research. On the communitarian end are the individuals who are the most dedicated to teaching. These teachers may not ever have harbored research interests, or they may have given up on these interests at relatively early career phases. Either way, the communitarian world presents itself as a place where these individuals have a home.

Notice also that this theoretic depiction spaces the worlds apart such that the elite and the communitarian worlds touch one another, albeit with little overlap. This spacing depicts the subset of communitarians who exhibit elite-like careers: they are highly active in research and adopt an orientation that is geared toward research opportunities, incentives, expectations, and symbols. At the same time, their exposure to these conditions is limited because of their less-conducive institutional location. Although the conditions of this world prevent its members from following elites to a close approximation, they allow for what I earlier called miniature versions of elite careers.

By the same token, some elites resemble communitarians, although this proves to be uncommon. When this does occur, elites have normally decelerated sharply in their careers. This most often happens at a mid-to-late career stage. These elites reorder priorities, they normally embrace teaching over research, and they usually adopt an "expressive" rather than an "achievement" orientation to work.

Empirical Perspective

The elites, pluralists, and communitarians of this study do not directly correspond to the spacing of worlds as presented in figure 1. As I explained in chapter 1, when I selected universities I "maximized variation" in careers by choosing institutions that fell at or near the top, middle, and bottom of a scale established by the National Research Council. Given the differences in rank, I anticipated that the sample would consist of three wholly separate worlds.

But narratives and other characteristics of scientists' careers demonstrate that these worlds are not spaced evenly apart. In this selection of schools, the pluralist world is closer to the elite than the communitarian is to either of the other two (see tables 2, 3, and 5.) This is most likely accounted for by a ranking of the pluralist school that landed it above the median but nevertheless closer to the center of the distribution than to the top. I represent the social order of the worlds—as they exist empirically—in the Venn diagram in figure 2.

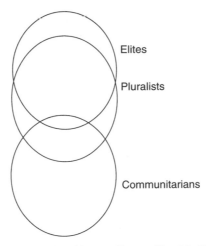

Figure 2   The Social Order of Science Careers (Empirical Formulation)

Notice how this sample of schools departs from how a set would be spaced on purely theoretic grounds. The diagram reflects the pattern here, in which pluralists bear, on average, greater resemblance to elites than to communitarians. Following the logic I have laid out, had I selected a different pluralist institution, one closer to the median ranking, the diagram would be more similar, if not parallel, to the theoretic formulation in figure 1.

One can postulate sociocultural distance and verify it empirically using a host of different samples. Where do the borders of Stanford, Arizona, and New Mexico State overlap, if at all? What about the University of Texas, Northeast Missouri State, the State University of New York at Buffalo, or any other set of institutions? The theoretic problem this issue raises consists in knowing when and why one institution is more elite, pluralist, or communitarian than another. The solution I propose is found only through comparison. University X is more communitarian than University Y because fewer members of University X resemble pluralists and still fewer resemble elites (in objective characteristics and in narratives). University A is more elite than University B because fewer members of University A resemble pluralists and still fewer resemble communitarians.

The comparative basis by which we identify institutions and their members underscores their location on a sociocultural plane. Institutions exist in a social space, their boundaries defined by their sociocultural similarities and differences. In principle, the entire population of higher education institutions can be located at some interval on the spectrum covering elite, pluralist, and communitarian types.

Once we locate a given institution, we can predict with high confidence what an average person's career will look like and how the career will be accounted for. In other words, people's career narratives have an ecology that speaks of the relationship between people and the constitutive elements of their work environments. Individuals follow a master narrative (or some branch thereof), highlighting the most salient features that distinctively characterize life in their world.

Ecology, Narrative, Self-Identity

As this study has sought to demonstrate, narrative is the key to understanding people in time and place. Narrative socially situates people, telling a story that meaningfully locates lives within the grand spectrum of life itself. Having a narrative and only a narrative is, I have argued, the best guide to identifying the context from which one speaks. Narrative informs us of the meanings that individuals impute to their lives, and those meanings have a contextual base—in this instance, elite, pluralist, or communitarian.

On one level, we have dealt with individual accounts. But we must remember that the narratives are shared, part of an ecology. Based on prevailing structural and cultural conditions, they take a form that distinguishes one group from another (cf. Czarniawska 1997). People "cool-out" in patterned ways, and they narrate their careers in ways that reflect these socially situated patterns. Even for those individuals whose narratives depart from the master form, we can pinpoint the world from which the narrators speak or, at a minimum, the worlds that they straddle, because the narrative itself exists within (and is a product of) a socially situated space.

These groups, like many others, consist of what some might call distinct "speech communities" (Mills 1940; Scott and Lyman 1968). Just as we encounter different languages in the world at large, so too do we hear people narrate their careers in systematically different fashions throughout the worlds of science (and other occupational worlds). To that end, narrative can operate not only as a means of claiming and crafting identity but also as a resource: those who inhabit distinct speech communities extend information and intelligence in differential ways, with significant life consequences (cf. Bruner 1996). As Harriet Zuckerman (1977) has found, the chances of winning a Nobel Prize increase exponentially when one works in the company of other prizewinners, not simply because there is a push for exemplary performance (though that is a real factor) but also because one has access to privileged information, codified in the narratives of those with whom one works. When you swim with Olympic swimmers, a scientist earlier observed, you begin to act (and talk) like one. The stories we tell ourselves and others—composed as those stories are of the ingredients of lives—become agents of stratification.[8] In application, narrative therefore exists as a prime cultural tool by which to understand people in context.

Science as Narrative

The larger significance of the profession's internal order rests with this study's principal theoretic concern: ambition. Social worlds inspire and shape ambition in different ways. In the sixty life stories disclosed in these pages, whether individually or collectively, scientists have discussed how they view their professional lives. In hearing their stories, we have developed a sense of what their lives are like, of what they find meaningful and significant, of how they orient an understanding of their passage. We have heard the ways in which their lives are similar, with each ordered by formative phases and transitions, and the ways in which their lives are radically different, with these phases and transitions experienced differently by different people and experienced not at all by others. We have entered their internal conversations, disputes, and reconciliations about where they see themselves standing or wanting to stand, where they have been, and where they see themselves headed.

How do we link ambition with the social order of science? Through narrative. Individuals use narrative to know and locate themselves and others within a complex array of types, turning points, and trajectories. As an organized occupation, science may be viewed as an array of narratives, each characterized by a script that individuals have followed or are attempting to follow. As I have explained, the scripts may take one of three dominant forms: of being, of becoming, or of something in between. Alternatively, the scripts may assume a form that constitutes a variation on one of these master themes, branching out from the root paradigm.

Who Shall I Be?

To a degree, narratives can be chosen, emphasizing the role of individuals as "free agents." People can portray themselves by highlighting how they want to know themselves and how they want to show themselves to others. In extremes, people can present themselves in ways that bear little or no resemblance to reality. Individuals embellish themselves to impress, persuade, or control others, to establish credibility, or to console themselves. They risk blowing their cover with those who know the facts or whose prior knowledge contradicts or calls into question the image being presented.[9]

But narratives are also constrained by the worlds in which people work. They are shaped by the sociocultural conditions that differentiate worlds. Narratives of becoming (elites) are found in worlds of plentiful opportunity, strong incentives, high expectations, and grand symbols. By contrast, narratives of being (communitarians) are found in worlds of comparatively meager opportunity, weak incentives, low expectations, and mild symbols. Narratives in between being and becoming, hybrids of the polarities (pluralists), are found in worlds in which the sociocultural conditions themselves are a combination of the above extremes.

Thus although an individual has choices in selecting a narrative, these choices are limited by the environment in which he or she operates. Environments vary in the restrictions they place on the narratives that individuals invoke. For example, certain elites may want to do something else before their careers have come to an appropriate end (at age sixty-five, for instance). But such individuals normally realize that the costs associated with a radical change in course are too high. Withdrawing from research in order to cultivate extracurricular interests entails troubling consequences, loss of esteem being foremost among them. The elite physicist who suddenly leaves research in order to play the guitar embarks on a path (and invokes a narrative) that colleagues denigrate.

The choices available to individuals may be set in terms of a continuum of proscriptiveness and permissiveness (cf. Parsons 1951). As indicated above, the roles that elites can perform in keeping with local standards are few. The elite world is, in other words, highly proscriptive, setting clear and comparatively inflexible parameters on exactly which roles count as legitimate and valued. The esteemed scientist is engaged and remains engaged over the career, although the specific forms of engagement may vary across life. The young scientist performs the role of the budding researcher, whereas the more advanced scientist may play more of a role as manager in addition to researcher—as a national spokesperson for the field, an administrator, an executive officer, a diplomat, or some other role.

Moving from the elite into the pluralist and communitarian worlds, roles become less proscriptive and more permissive. Different kinds of careers become increasingly possible. By the same token, we find more diverse narratives that reflect heterogeneity in outlook, orientation, and ambition. Thus communitarians talked about turning away from science, entertaining and often acting on entrepreneurial, recreational, familial, and leisurely interests, and doing so with few social costs. In fact extraprofessional achievements, whether in business, recreation, or family, are often esteemed by fellow communitarians. Most follow the old adage—"don't take work too seriously"—and community gets built on people's livelihoods in and out of work.

Elites are more inclined to judge others on the basis of their latest scholarly work. Whether one built a house, became an oil magnate, was an all-star tennis player, or championed twelve promising children is secondary or altogether irrelevant. The worlds of science—elite, pluralist, and communitarian—exist, therefore, at different positions on a moral playing field.

Lives in Science

If we view science—indeed any occupation—as an array of narratives, then ambition itself tells a story. Ambition may be understood as the internalization of a narrative. For young scientists, the narrative may consist of a scenario: the rising star performing magnificent feats, receiving great honors, attempting to secure a place in history. For older scientists, the narrative may consist less of a scenario and more of a summing-up and preservation of a life constituted in part by specific attainments. A life marked by successes, however many or few, needs to be protected from diminishment or lapses in the collective memory.[10] Older scientists may invoke the narrative of becoming, but they are aware that any additional status increments will likely be small.[11] Nevertheless, the rhetoric of becoming (or having become) plays an important part in the process of self-preservation. Thus while the form of the narrative remains more or less the same over the life course—emphasizing progression—the functions of the narrative change, serving at the start as an image of a life to be lived and later as a statement of a life to be safeguarded.

Greatness sets itself apart from all other quests because it speaks not only of those who keep chasing it but of all who in their journeys have unavoidably been tempted by its promises, tantalized by its illusions. Greatness possesses all those people—in and out of science—whose peak performances have brought them closer to their goal and at the same time raised the stakes in a never-fully-satisfiable affair. But more generally, greatness molds the careers of all because it is with such illusions of grandeur that individuals forge and often sustain their lives in science, in architecture, in music, in medicine, in law, in politics, in sports. Ambition plays the lead role in all endeavors in which people honorably seek to in-

fluence and advance civilization, for what would anything less than ambition do for such exalted aims?

Those who turn the call away nevertheless are aware of the kind of life
that ambition prescribes. Greatness can be seen as so daunting, so insur-
mountable, that individuals are easily doused. When people face the reality of its elusiveness for the final time, they must knowingly put
ambition aside and settle on smaller sights, emphasize the benefits of the
present, or resign themselves to defeat altogether.

For those who follow ambition's lead, though, the career becomes (or
remains) an anchor, the major force with which people organize their activity. Work floods over into other parts of life. This dramatic—and perhaps overtaxed—image of ambition as a vital lifeline is underscored by
scientists' accounts of its all-embracing meaning.

> To have ambition is to strive to attain certain things which are not easy
> to attain. That's my ambition. When you are very young, if you're suc-
> cessful, then the very fact that you're successful drives you forward
> because of the excitement in that. You end up reading biographies of
> the great men, and you want to emulate them. Work is an essential ele-
> ment, and being a very successful scientist has that stress. If you are
> not happy and successful at work, you are not going to be happy and
> successful, period.[12]

> Physics is the central focus of my life. It's what I think about when I
> go to bed, and it's what I do here during the day. I spend very many
> evenings thinking about physics. My will to succeed comes from a
> recognition that it makes for a superior approach. If you play to win,
> it's more fun. It's boring to play sports if you don't play to win. And if
> you're going to do science and you don't do it with the intention of
> succeeding or uncovering something new, you're just doing it in a
> second-rate way.[13]

> I am conscious of this scale [of scientific greatness] and where I fit. In
> what little time I have I usually read biographies or histories of twen-
> tieth-century physics and mathematics, and I am very conscious of
> the history, not at the level of an historian but at the level of more than
> just an amateur. I really care about the history and the context and the
> role that twentieth-century science has played in shaping civilization,
> and I see myself as a tiny fragment of that tapestry. I really care about
> that whole tapestry, and though I don't want to be the best or the most
> famous, there is a certain level I'm aiming for. It's somewhere near
> "hotshot," because I think that's the best I can do. I want my event.
> That's the hardest concept to get across, I think. Ideas go with names

in physics, very, very clearly. I want mine. I want people to know it was mine.[14]

As these accounts show, ambition organizes action, effort, and energy toward noble causes. But these accounts also point to broader themes that portray people's lives playing out through the complex vicissitudes of time and place. The dream of heroic greatness is what propels individuals and sustains their ambition. But greatness is almost never achieved. Accomplishing one great thing is simply an occasion to realize that one must accomplish something greater. Most people are not Nobel Prize winners; most people aren't prizewinners period.

All educational institutions, along with many occupations, raise the expectations of the young. A good many of them do this deliberately, as a complex process of induction and socialization. And a good many succeed, for what comes from people who have neither thirst nor hunger? Schooling creates an illusion of heroic individual performance. People aspire to greatness—making their mark, assuming their place in history, taking their tailored seat among the pantheon. Their passions are stirred in the shadows of gods whom they hope one day they will sit beside (or at least not too many rows behind): the Albert Einsteins, the Enrico Fermis, the Albert Schweitzers, the Pablo Picassos, the Michael Jordans. The pantheon, whether of physics or of football, enshrines the great figures who have passed before, setting a standard while simultaneously propagating heroic myths for those who follow. As a social historian has observed, "Without dreams of achievement, and ambition to fuel them, life itself might as well be over" (Epstein 1980, 8).

The rhetoric of heroic accomplishment is forever sustained. The pantheon endures, occasionally welcoming new heroes. The wider culture continues to value dreams and a life course in which ambition makes those dreams come true.

> I'll give to you a house of mirrors,
> A thousand eyes they belong to you;
> A labyrinth of wild roses,
> I know you'll find your own way through.
> Wait a while and you'll grow older,
> But never mind what the old folks say;
> Just keep an angel on your shoulder,
> And never throw your dreams away;
> For they might save your life one day.[15]

But even though our culture sustains a romantic vision, in the end the situation, for most of us, is drastically transformed. Voices alter in tenor; visions change in focus. Needless to say, people are pained to let go of their dreams. The reward system of science—the reward system of any art or science—must be looked at not only as a system that provides incentives but also as a system that preserves the idea of individual heroism in the midst of overwhelming odds.

The ordering of careers into never-fully-realizable dreams suggests that the achievement of greatness may be no more worthy of honor than the quest itself. What may not be found with great careers, if only because they are rare, can be found in great poetry, which often speaks to the more common lot. Those who seek to change the world must ultimately be "Bred to a harder thing Than triumph." "Be secret and exult," Yeats consoles; accept defeat quietly and take satisfaction in that achieved.[16] It is therefore easy to understand why scientists—elite, pluralist, and communitarian—place the remembrance of their lives more often in humble than in majestic terms. Very few achieve fame, and even when it is won, great recognition often eludes people in their own lifetimes. The moral career unites and divides. It is what scientists hold in common, a lifeline to which all cling, moving together and apart.

APPENDIX A

Interview Questions

Interview of Scientists

This is a study about the aspirations of academic scientists. The questions I would like to talk about deal with one's individual identity and how that identity has unfolded over time. Some of the things I will discuss ask you to reflect upon yourself and often involve making personal judgments that will touch on various professional and related personal topics. Your participation in this study is strictly confidential. Interviews are normally tape-recorded, and this simply provides for accurately keeping track of information. Subsequently the tape will be destroyed. Your participation in this study is important. However, should you at any time wish to stop, you may do so without prejudice to you, and at any time you should feel free to ask me questions concerning the interview or the study. May we begin?

A. Location in the Division of Scientific Labor

1. Can you describe the type of work you do?
2. To what extent is your work collaborative?
3. [If collaborative] How large are the collaborative teams on which you work?

B. Construction of Personal Histories and Personal Identities

1. What aspirations did you have as a graduate student?
   Probe: What did you want to attain?
2. In everyone's career there are "roads not taken"—different avenues you might have followed. What have been the ones for you?
3. What consequences have these outcomes had on your career?
4. How did you come to arrive at this university?
5. You were a graduate student at _____. Is this the type of university where you wanted to end up?

6. How have your aspirations unfolded since being a graduate student?
7. How has being at this university affected your career?
   Probe: How has this university constrained your career? How has it helped your career?

C. *Generalized Definitions of Success Ladders*

1. What do you associate with a "successful" career in physics?
2. What do you think are the most important qualities needed to be successful at the type of work you do?
3. What does *ultimate* success mean to people working here?
4. Is there an understanding of a *minimum* needed in order to maintain respect among people here?
5. Is there an understanding of a *failed* career among colleagues here?
6. Taking your colleagues in this department, how would you say their success varies?
   Probe: Have they advanced at the same rate?
7. Where do you place yourself among that variety?

D. *Conceptions of Future and Immortalized Selves*

1. What do you dream about in terms of your career?
2. What ultimate thing would you like to achieve?
3. How do you envision yourself at the end of your career?
4. How would you like to be remembered by your colleagues?
5. What about your life do you think will outlive you?

E. *Ambition*

1. Would you say that you are ambitious?
   Probe: Would you say that you have a strong will to succeed?
2a. [If yes to 1] What is your ambition?
2b. [If no to 1] Would you say that you have a strong will to succeed?
3. Where does your ambition come from?
4. What role do you think ambition plays in your life?

F. *Self-Doubt/Self-Fragmentation*

1. What would you like to be better at?
2. Has there been a significant time when things really did not go the way you wanted them to?
3. What major doubts have you had about yourself?

4. Have there been times when you felt that you let yourself down?
   Probe: Have you ever felt disappointed in yourself?
5. Has there been some inner conflict or turmoil that you have sought to understand in your life?

We are near the end of the questions I have.

6. I would finally like to ask about something you are most proud of. What stands out as something that has left a strong positive impression on you?

APPENDIX B

Postinterview Questionnaire

Study of Science Careers

Department of Sociology
The University of Chicago
Chicago, Illinois 60637

Information provided on this form is strictly confidential and will be used for research purposes only. At all times your identity will remain anonymous.

1. Please list *up to four* subfields of physics with which you most closely identify.

   1. _____

   2. _____ (*if appropriate*)

   3. _____ (*if appropriate*)

   4. _____ (*if appropriate*)

2. In which professional organizations are you *currently* a member? (*Check all that apply.*)
   1. ( ) American Physical Society
   2. ( ) American Association of Physics Teachers
   3. ( ) Optical Society of America
   4. ( ) IEEE Lasers and Electro-Optics Society
   5. ( ) Society of Rheology
   6. ( ) American Crystallographic Association
   7. ( ) American Astronomical Society
   8. ( ) American Association of Physicists in Medicine
   9. ( ) American Vacuum Society
   10. ( ) American Astronautical Society
   11. ( ) American Meteorological Society
   12. ( ) American Nuclear Society
   13. ( ) Health Physics Society
   14. ( ) Acoustical Society of America
   15. ( ) American Association for the Advancement of Science
   16. ( ) **None of the above**
       ( ) **OTHER** (*specify*)

3. List all employment, appointments, or other positions, *aside from those listed on your curriculum vitae,* held between *leaving college* and *uninterrupted, full-time* professorial employment.

POSITION/TITLE EMPLOYER CITY, STATE DATES OF EMPLOYMENT

1. _____

2. _____

3. _____

4. _____

5. _____

4. Have you attended a *national* meeting of the *APS*?
   1. ( ) Yes
   2. ( ) No

5. If *YES*, when was the last *national* meeting of the *APS* that you attended?
   19___.

6. How often do you attend the *national* meeting of the *APS*?
   1. ( ) every year
   2. ( ) about every other year
   3. ( ) about every 3–4 years
   4. ( ) about every 5 years
   5. ( ) less than every 5 years
   6. ( ) never

7. Have you attended a *section meeting* of the *APS*?
   1. ( ) Yes
   2. ( ) No

8. If *YES*, when was the last *section meeting* that you attended?
   19___.

9. How often do you attend *section meetings* of the *APS*?
   1. ( ) every year
   2. ( ) about every other year
   3. ( ) about every 3–4 years
   4. ( ) about every 5 years
   5. ( ) less than every 5 years
   6. ( ) never

10. Have you attended a *national* meeting of the *AAPT*?
    1. ( ) Yes
    2. ( ) No

11. If *YES*, when was the last *national* meeting of the *AAPT* that you attended?
    19___.

12. How often do you attend the *national* meeting of the *AAPT*?
    1. ( ) every year
    2. ( ) about every other year
    3. ( ) about every 3–4 years
    4. ( ) about every 5 years
    5. ( ) less than every 5 years
    6. ( ) never

13. Do you attend other *national, international,* or *regional* physics meetings?
    1. ( ) Yes
    2. ( ) No

14. If *YES,* list the meetings and indicate how regularly you attend them.

MEETING	every year	about every other year	about every 3–4 years	about every 5 years	less than every 5 years
1. _____	( )	( )	( )	( )	( )
2. _____	( )	( )	( )	( )	( )
3. _____	( )	( )	( )	( )	( )
4. _____	( )	( )	( )	( )	( )
5. _____	( )	( )	( )	( )	( )

15. To which professional journals do you currently subscribe?
    1. ( ) *Physical Review*
    2. ( ) *Physical Review Letters*
    3. ( ) *Physics Letters*
    4. ( ) *Bulletin of the American Physical Society*
    5. ( ) *Journal of Physics*
    6. ( ) *Chemical Physics Letters*
    7. ( ) *Reviews of Modern Physics*
    8. ( ) *Optics Letters*
    9. ( ) *Nuclear Physics*
    10. ( ) *Journal of Physics and Chemistry of Solids*
    11. ( ) *Journal of Computational Physics*
    12. ( ) *Journal of Contemporary Physics*
    13. ( ) *Journal of Experimental and Theoretical Physics*
    14. ( ) *Journal of Low Temperature Physics*
    15. ( ) *Journal of Mathematical Physics*
    16. ( ) *Journal of Mathematics and Physics*
    17. ( ) *Journal of Plasma Physics*
    18. ( ) *Journal of Polymer Physics*
    19. ( ) *Journal of Earth and Space Physics*
    20. ( ) *Journal of Statistical Physics*
    21. ( ) *Journal of Mechanics and Physics of Solids*
    22. ( ) *Applied Physics Letters*
    23. ( ) *Applied Physics*
    24. ( ) *The American Journal of Physics*
    25. ( ) *The Physics Teacher*
    26. ( ) *Journal of Computers in Mathematics and Science Teaching*
    27. ( ) *Science* (AAAS)
    28. ( ) **None of the above**
       ( ) **OTHER** *(specify below)*
    _____

16. Name and describe your father's occupation.

    Job title or position: _____

    What does/did he do? _____

    _____

17. Name and describe your mother's occupation.

Job title or position: _____

What does/did she do? _____

_____

18. In what year were you born? 1 9___.

19. What is your current marital status?
  1. ( ) Married
  2. ( ) Separated
  3. ( ) Divorced
  4. ( ) Single

20. Do you have children?
  1. ( ) Yes
  2. ( ) No

21. If *YES*, please list the year(s) of birth of your child(ren).
  1. 1 9 ___   6. 1 9 ___
  2. 1 9 ___   7. 1 9 ___
  3. 1 9 ___   8. 1 9 ___
  4. 1 9 ___   9. 1 9 ___
  5. 1 9 ___   10. 1 9 ___

22. Please indicate your past year's *nine-month base university* salary. (Do *not* include nonuniversity income, e.g., research grants, consulting fees, royalties, honoraria, etc.)
  1. ( ) less than $20,000   7. ( ) $70,000–$79,999
  2. ( ) $20,000–$29,999   8. ( ) $80,000–$89,999
  3. ( ) $30,000–$39,999   9. ( ) $90,000–$99,999
  4. ( ) $40,000–$49,999   10. ( ) $100,000–$109,999
  5. ( ) $50,000–$59,999   11. ( ) $110,000–$119,999
  6. ( ) $60,000–$69,999   12. ( ) greater than $120,000

23. Please indicate your current religious preference.
  1. ( ) None
  2. ( ) Baptist
  3. ( ) Buddhist
  4. ( ) Congregational
  5. ( ) Eastern Orthodox
  6. ( ) Episcopal
  7. ( ) Hindu
  8. ( ) Islamic
  9. ( ) Jewish
  10. ( ) LDS (Mormon)
  11. ( ) Lutheran
  12. ( ) Methodist
  13. ( ) Moslem
  14. ( ) Presbyterian
  15. ( ) Quaker
  16. ( ) Roman Catholic
  17. ( ) Seventh-Day Adventist
  18. ( ) Tao
  19. ( ) OTHER (*specify below*)

_____

24. In the past year, about how often have you attended religious services?
    1. ( ) more than once a week
    2. ( ) about once a week
    3. ( ) two or three times a month
    4. ( ) about once a month
    5. ( ) several times a year or less
    6. ( ) not at all

25. What is your racial/ethnic background?
    1. ( ) White/Caucasian
    2. ( ) Black/African-American
    3. ( ) Chinese-American
    4. ( ) Japanese-American
    5. ( ) Indian
    6. ( ) Israeli
    7. ( ) Hispanic
    8. ( ) OTHER *(specify below)*

    _____

26. In what country were you born? _____

27. In what country or countries do you have citizenship?

    _____   _____

APPENDIX C

Departmental Questionnaire

Study of Science Careers

Department of Sociology
The University of Chicago
Chicago, Illinois 60637

Information provided on this form is strictly confidential and will be used for research purposes only. At all times, including in any published work, your identity and the identity of your institution will remain anonymous.

Please answer the following questions as completely and as accurately as possible. Kindly return the completed form in the enclosed self-addressed, stamped envelope. Your assistance in this research effort is greatly appreciated.

I. Teaching

1. What is the normal yearly teaching load for the average faculty member of your department?

   ( ) 1–2 courses   ( ) 4–5 courses
   ( ) 2–3 courses   ( ) 5–6 courses
   ( ) 3–4 courses   ( ) More than 6 courses

2. For the average faculty member in your department, what percentage of teaching in a given year involves general undergraduate courses?

   ( ) 75–100%
   ( ) 50–74%
   ( ) 25–49%
   ( ) Less than 25%

3. What percentage of faculty members in your department teach mostly courses at the graduate level?

   ( ) 75–100%
   ( ) 50–74%
   ( ) 25–49%
   ( ) Less than 25%

4. Does your department have provisions for leaves from teaching?

   ( ) Yes   ( ) No

5. *If yes to Question 4,* explain briefly the conditions under which a faculty member can take a teaching leave.

6. *If yes to Question 4,* how often can leaves from teaching be taken?

II. Resources

1. What is the amount of *annual federal support* to your department?

2. What is the amount of *annual nonfederal support* to your department?

3. What is the overall annual operating budget of your department?

4. Taking all physics departments in the United States, how would you characterize the present *equipment and research facilities* in your department?
   ( ) among the very best
   ( ) very good
   ( ) average
   ( ) fair
   ( ) relatively poor
   ( ) nonexistent

5. What would you take to be the greatest problem(s) with your present equipment and research facilities? (Check all that apply.)
   ( ) Generally, they're old.
   ( ) Generally, they're limited.
   ( ) Generally, they do not allow for major research.
   ( ) They serve a relatively small fraction of the department faculty.
   ( ) Other (indicate below):
   _____
   ( ) None of the above applies.

6. Do your new assistant professors receive university or department start-up funds or a similar allowance?
   ( ) Yes   ( ) No

7. *If yes to Question 6,* what is the amount?

8. *If yes to Question 6,* do these funds have restricted professional uses?
   ( ) Yes   ( ) No

9. *If yes to Question 8,* explain what the funds may be used for.

10. Do faculty members in your department have access to *departmental or university* funds for research?
   ( ) Yes   ( ) No

11. *If yes to Question 10,* indicate the type(s) and amount of fund(s).

   Type: _____   Amount: _____

   _____   _____

   _____   _____

12. Is each faculty member in your department covered for travel/conference expenditures with departmental or university funds?
   ( ) Yes   ( ) No

13. *If yes to Question 12,* for how many trips/conferences per year? (Alternatively, explain the provisions.)

14. Roughly how many employed *technical staff* are on hand for faculty members in your department? (Exclude postdocs and graduate assistants.)
    ( ) 1–5   ( ) 16–20
    ( ) 6–10   ( ) 21–25
    ( ) 11–15 ( ) More than 25

15. In a given year, roughly how many *postdoctoral researchers* have appointments in your department?
    ( ) 1–5   ( ) 16–20
    ( ) 6–10   ( ) 21–25
    ( ) 11–15 ( ) More than 25

16. Roughly what percentage of faculty members in your department have *graduate research assistants*?
    ( ) 0%
    ( ) 1–25%
    ( ) 26–50%
    ( ) 51–75%
    ( ) 76–100%

17. Taking all the faculty members who do *not* have graduate research assistants, what would you say is the single most important reason? (Check only one.)
    ( ) They do not need r.a.'s because work is theoretical.
    ( ) They do not need r.a.'s because not active in research.
    ( ) Pool of r.a.'s is too small.
    ( ) Talent pool of r.a.'s is too weak.
    ( ) Other (indicate below):

_____

18. Using your best judgment, estimate the number of outside speakers who pass through your department in an average academic year.
    ( ) None   ( ) 16–20
    ( ) 1–5   ( ) 21–25
    ( ) 6–10   ( ) 26–30
    ( ) 11–15 ( ) More than 30

III. Personnel Policy

1. Is it customary for your department to solicit outside letters for review of candidates being evaluated for tenure?
    ( ) Yes   ( ) No

2. Was this a customary practice for your department ten years ago?
    ( ) Yes   ( ) No

3. Twenty-five years ago?
    ( ) Yes   ( ) No

4. As best as possible, indicate the extent to which teaching performance factors into promotion decisions in your department.
    ( ) A lot
    ( ) Somewhat
    ( ) A little
    ( ) Practically not at all
    ( ) Not at all

IV. Additional Comments

Please make any additional comments you wish to make, including those that you feel would help inform any of your responses above.

Thank you for taking the time to complete this form.
Please return in the enclosed envelope.

Notes

Introduction

1. See Cogan 1953; Etzioni 1969; Greenwood 1957; Marshall 1939; Parsons 1949; Wilensky 1964. For more current treatments of professions, including several that deemphasize trait-based approaches, see Abbott 1988; Brint 1994; Freidson 1986; Halliday 1987; Larson 1978. Despite being published many years ago, Johnson's account (1972) remains an excellent conceptual overview of the professions field.

2. "Life course" is often used interchangeably, but should not be confused, with the similar terms "life cycle" and "life span." Life cycle transcends the single lifetime of an individual and invokes the sense of repeating generational processes, driven by reproduction, that occur in natural populations. Life span refers loosely to the period from birth to death. For elaboration on these terms and their usage, see Elder (1992) and O'Rand and Krecker (1990).

3. Aristotle, *Nicomachean Ethics*, lines 1125b–1151b.

4. Aristotle's reading of ambition suggests that its excesses are as costly as its deficits. The term "ambition" clearly has political connotations, particularly when its excesses are associated with greed or self-interest (e.g., political ambition). The Latin root of ambition—*ambitio*—means to search around for votes, further highlighting the political overtones of the term. Deficits of ambition similarly arouse social concern (e.g., parents' view of a child or professionals' view of a colleague), as exemplified in the culturally standardized appraisal "He lacks ambition." Aristotle's concern with identifying the mean of ambition's excesses and deficits is both timely and timeless. Portions of society are characterized as ruthlessly self-interested, whereas others appear to lack any productive motivation whatsoever. There are social as well as economic consequences to this behavior. But because the concern of Aristotle—and of those who share it—involves an inherent search for a moral understanding, the politics of ambition (how much is right, how much is wrong) is best debated by philosophers and theologians, not sociologists.

5. *The Confessions of St. Augustine*, book 11, chapter 11, "Past, Present, and Future."

6. The line of work known as "life course studies" spans several disciplines and interdisciplinary fields: sociology, psychology, anthropology, biology, hu-

man development, gerontology. My use of work in the field will necessarily be selective. I will be concerned primarily with those sociological aspects of life course that inquire about the link between biography and environment. Several excellent overviews of the sociology of the life course include Elder (1975, 1985, 1992, 1994, 1998), Featherman (1983), Hagestad (1990), Riley (1988), and Riley, Foner, and Waring (1988). Other disciplinary perspectives can be found in Binstock and George (1990) and Sorensen, Weinert, and Sherrod (1986).

7. Related work has considered how both the life course and people's transitions through it are viewed differently at different historical periods in time, due to prevailing structural arrangements in institutions such as the economy and the family (Buchmann 1989; T. Cole 1992; Hareven 1978, 1986; Riley 1986; Riley, Johnson, and Foner 1972). In a similar vein, Mayer and Schoepflin (1989; Mayer 1986a, 1986b) have put forth a macrotheoretic perspective that accounts for the formation and experience of the life course in terms of constraints and regulations imposed by the state.

8. See also the pointed response to his critique in Baltes and Nesselroade (1984) and Dannefer's rejoinder (1984b).

9. See Kohli (1986) for an example of how workers in German industry construct biographies in light of social forces that stem from the organization of work.

10. For an overview of the sociology of science, including developments in research, see Zuckerman (1988) and compare this with the similar overview published a quarter-century before (Kaplan 1964).

11. Merton's article "The Normative Structure of Science" was published in 1942; see Merton 1973b. For related work on the "ethos" of science, see Ben-David's many contributions, compiled in a posthumous anthology (1991, especially parts IV–VII). For an ever-principled discussion of the ethos of higher learning broadly construed, see Shils (1983).

12. The norm of *communism* (also known as "communalism") holds that knowledge must be shared, not kept secret. The norm of *disinterestedness* holds that the motives and actual conduct of science should be driven without personal bias. The norm of *organized skepticism* holds that scientific judgments are to be held until all necessary evidence is on hand. Each norm of science, in its own way, is seen to exist in order to preserve the stable functioning of science and to serve the chief goal of science: the extension of certified knowledge.

13. Clark (1987a) provides an extensive treatment of many variables that factor into the way in which the American professoriat is socially organized, including distinctions in institutional culture, competing bases of organizational authority, and the differentiation of academic career lines. Moreover, the analysis includes narrative accounts from diverse individual practitioners. The treatment, though, is set in static terms rather than the dynamic terms that a life course perspective offers. Professional self-identity—how it forms and evolves as a function of the context of work—thus falls outside the scope of Clark's mission.

14. S. Cole (1992) provides an adjudication of the constructionist and more normative-Mertonian stances. Shapin (1995) provides an account of the development of constructionist studies of science, including the controversies that constructionist study has generated vis-à-vis the older structural-functional line of inquiry. See Zuckerman (1988) for a comprehensive treatment of the debates.

15. According to the latest survey, there are 387,740 employed scientists in the United States who hold a doctor's degree. Of these, 51.0 percent are employed in academia, 26.1 percent in industry, and 11.0 percent in federal, state, and local government. The remainder are employed in other miscellaneous sectors (National Science Foundation 1993).

16. As I have discussed, I will be primarily concerned with variation across contexts of work (place) and across time. To this end, the "units of analysis" are the *academic department*, the *university*, and the *individual*. Within-field variation, that is, differences among particular types of physicists, will receive only brief consideration at points where relevant. Although there are sure to be some interesting within-field distinctions, the range of variance in identity construction is likely to be greater along contextual and temporal lines, thereby yielding a larger social story to tell (and one with a more general appeal). Those in search of a *systematic* analysis of within-field distinctions among physicists should see already existing sources: Krieger (1992) and Traweek (1988).

17. My discussion of this point follows Suttles (1995), on whose insight I will rely heavily in the next several pages.

18. Consult any one of the number of physics textbooks or biographical accounts in which physicists describe the technical aspects of their work.

19. Readers should be informed that definitions of science on objective, cognitive grounds are taken by some to be contestable (see the discussion above, "Echoes of Time and Place"; for a broader statement on the issue, see Gieryn forthcoming). For these purposes, however, I have no reason to ask my subjects what they "believe" science to be (i.e., how they "define" it), since I am dealing with the construction of the career, not knowledge. In doing so, we begin by accepting a conventional definition of the occupation we call "science" (see Zuckerman 1988, 513).

20. The modern concern with science and the preeminence of scientists may be seen to have its roots in what vintage social theorists knew as the "law of three stages," first articulated by Saint-Simon (1760–1825) and later developed by Auguste Comte (1798–1857), the latter credited with coining the term "sociology." The "law of three stages" consists of a theory of historical progress and a framework for interpreting all human thought. The three stages—each referring to a historical time and accompanying modes of thought—include the *theological*, the *metaphysical*, and the *positive.*

In the theological stage, we think anthropomorphically, providing explanations for phenomena that consist of qualities like our own. Society is ruled by priests and monarchy, to whom we turn for guidance and who in turn invoke demons and divinities to explain our world.

In the metaphysical stage, knowledge is speculative. It is associated with a negative historical period of social criticism, rebellion, and political upheaval.

In the positive stage—the most mature state of thought and being—a new framework for life is ushered in only when scientists resolve to think solely in terms of lawful relations between observed conditions in the empirical world. As Comte believed, the positive stage began a time when scientific, rather than speculative, knowledge—and scientists, rather than priests—could be used to organize society in harmonious ways, representing above all a rational path toward systematically improving the human condition.

For elaboration on the development of Comte's thought and the ideas that flowed from it, see Levine (1995, especially 160–172).

21. Institutions have now taken to supporting and cultivating the biography industry, themselves becoming a player in the production of cultural legends. The Alfred P. Sloan Foundation alone has overseen the release of well over a dozen works, and other organizations are similarly engaged. For example, the American Institute of Physics publishes the series "Masters of Modern Science." Anthologies of scientific biography also are in plentiful supply, each account further dramatizing the way in which superlative lives in science mythify the ideal career. Such anthologies include the well-established *American Men and Women of Science,* first compiled in 1906 and now in its nineteenth edition.

22. For being so widely diffused throughout the culture, few if any social treatments exist that account for "ambition for greatness." Reflecting the widespread temptation to think about ambition and achievement as purely functions of the individual, some notable works take up the topic of greatness. Simonton (1994, also 1988), in his search for the answer to who makes history and why, provides a psychological account of people who fit the profile of "greatness." He is concerned with personality traits that "predispose" certain people to become great figures. Ludwig (1995), in his search to resolve the "creativity and madness controversy," is similarly concerned with ways in which individual traits may be "predictive" of greatness. He pays special attention to the link between creativity and mental illness.

23. The "exemplary professional passage" is of course a historical construction, for in many cases—Van Gogh, Mozart, Newton—few at the time thought the life to be extraordinary. Indeed at the time, the figure may well have been viewed as a maverick or may have gone wholly unnoticed (see DeNora 1995; Heinich 1996).

Chapter 1

1. All of the interviews were transcribed by the author. They constitute over seventy hours of interview tape and over twelve hundred single-spaced pages of transcript.

2. Had my selection of departments been based on any one of the other five or on all of the criteria combined, the selection of schools would have been little or no different from those in the present study. Each of the six quality measures, including scholarly reputability, are highly intercorrelated (see Jones, Lindzey, and Coggeshall 1982, Table 9.3, p. 166). Distinguished departments tend to have large enrollments with successful graduate students (i.e., short time to Ph.D. and postgraduate employment), large and effective libraries, abundant research support, and research-productive faculty members. Conversely, less distinguished departments tend to be weaker on each of these counts.

3. Two of the three schools grouped at the bottom were not officially ranked in the 1982 NRC study. In one case, an institution lacked a sufficient number of Ph.D graduates (at least ten) over a specified two-year period to be eligible for ranking. In the second case, the institution was not ranked because it no longer had an established Ph.D. program in physics, although it continued to maintain a master's program. Though not ranked in the NRC study, both of these departments are listed in the American Institute of Physics guide to graduate programs (American

Institute of Physics 1993). Given the general similarity in program size and the other departmental and institutional characteristics that these departments share with the one ranked at the bottom, they would probably occupy similar ranking locations.

4. A follow-up to the 1982 NRC study was conducted roughly a decade later. The results of the follow-up study were reported in *Research-Doctorate Programs in the United States: Continuity and Change* (Goldberger, Maher, and Flattau 1995) and received wide media attention. None of the departments selected for this study from the 1982 rankings had marked changes in their standing. All continued to hold rank at or near the top, middle, or bottom of the scales used to assess quality, including the scholarly reputability measure. Although none of the institutions changed in ways that would challenge their being characterized at the top, middle, or bottom of a scale, the middle institution in this study showed notable change: it improved, drawing nearer to the top schools than to the bottom ones. I will discuss the consequences of this shift in later chapters.

5. The sociological use of "world" as a metaphor to situate individuals and their roles in society is not unique to this study but would appear to be unique among studies of scientists and the academic profession generally (to my knowledge, the only notable exception is Clark [1987a]). The metaphor has been adopted by others to gain intimate knowledge about what it is like to be a part of a social world, what social conventions constitute social worlds and set them morally apart from others, and how people create meaning about themselves vis-à-vis their membership and nonmembership in social worlds. For example, Becker (1982) speaks of "art worlds." We learn how art is socially organized as an institution and what it is like to be a part of any one of four worlds of art: the world of integrated professionals, the world of mavericks, the world of folk artists, and the world of naive artists.

The long line of community studies in sociology has frequently made use of social world as a metaphor, implicitly or explicitly, to account for the differentiation, and the inner workings, of urban and rural enclaves. In Suttles (1968) we acquire a deep understanding of social worlds of the city: we learn what constitutes the worlds of Italians, Mexicans, Puerto Ricans, and blacks and the conditions under which those worlds remain separate, collide, and sometimes work together.

Strauss (1978, 1993; also Clarke 1990, 1991) has suggested that social world as metaphor represents the most promising way to understand nothing short of all society, its groupings and processes.

Two major concerns distinguish the social world approach. First, it shows an explicit concern for *social ecology* by locating and attempting to account for behavior in terms of environment or sociocultural milieu. Second, and flowing from the first, the social world approach is explicitly concerned with the *social-interactive basis* on which reality and meaning get created and framed. At root, the approach moves from the "ground up," seeing how macro-orders are established from micro-processes and in turn seeing how macro-forces effect micro-functioning.

6. For instance, Caplow and McGee (1958) speak of "major league," "minor league," and "bush league" departments. The widely used Carnegie Classification System (Carnegie Commission on Higher Education 1994, 1973), though value-neutral and adequate for several purposes, does not discriminate among institutions and thus is not effective here. For example, both Harvard University

and the University of Kentucky are classified as "Research I Institutions." Though I have no doubt that both of these schools are excellent in many ways, I suspect that the structures and cultures of these schools differ on a number of dimensions that the Carnegie Classification System fails to capture but that bear directly on our concern with career and self-identity.

7. The collective identities of the worlds are derived empirically—that is, they follow patterns that will be the basis of all that follows. Collective identities are introduced at the beginning merely to speak of patterns intelligibly.

8. For this reason, my use of "elite" is set apart from "elitist." Elite refers to a status position, elitist to a *belief* that one is part of a priviliged group. Not all elites are elitist; we thus keep to the proper term "elite."

9. I will use the term "world" to refer to both academic departments and the universities in which they are housed, but mainly the latter. This of course raises the question of how to think about contradictions in status between universities and their departments. For example, an elite department can sometimes be found in a pluralist institution and a pluralist department in an elite institution. In this sample, departmental and institutional status are congruent: elite departments are part of elite universities, and so on.

In terms of a more general system of classification, we may say that elite institutions are those that have a majority of elite departments; pluralist institutions are those that have a mix of department orientations; and communitarian institutions are those that have a majority of communitarian departments. All three institutional types inevitably contain *some* variety, as one would reasonably expect, though the range and the nature of variety differ from type to type. More will be said on these matters in chapter 6.

10. Differences in the number of schools that make up each type stem from my effort to interview equal numbers of people across the three types. Communitarian departments are typically small; to reach the quota of communitarian respondents, I thus had to travel to more such schools.

11. For an assessment of Thomas and Znaniecki's seminal contribution to the development of sociological theory and method, consult Bulmer (1984, chapter 4). See also Coser (1971, 511–559) and Cahill, Fine, and Grant (1995).

12. On biographical approaches to the study of social life, see Denzin (1989) and Plummer (1983). For an account of major current substantive trends in the narrative tradition, refer to Bertaux and Kohli (1984). For general background on narrative analysis in sociology, consult Orbuch (1997), Richardson (1990), Riessman (1993), and Scott and Lyman (1968). For background on the sociological use of life histories in particular, see Maines (1992).

13. This is also true for many other fields. For a starkly different view of market conditions as they existed in the 1950s, see Caplow and McGee (1958).

Chapter 2

1. For general background, the reader may want to consult several older, now-classic studies that have examined the "character" of academic life, each focusing in its own way on various social-organizational features of the profession and the university or college. Wilson's (1942) study represents the first full-scale empirical treatment of the American professoriat as an organized profession. He is pri-

marily concerned with the form and function of hierarchy in the profession and with the profession's social status. Caplow and McGee (1958) address the functioning of the academic labor market, including processes of recruitment and selection. Lazarsfeld and Thielens (1958) address the political persuasions of academics and the midcentury attacks on academic freedom. Jencks and Riesman (1968) account for what they saw as the rise in power among academics, power exercised both in and outside of the academy, and explain the social consequences, for good and ill, of a meritocratically based system of higher education. Blau (1973) offers a structural view on the organization of academic work; he is concerned with the operations of the university and with what conditions of practice foster and inhibit the work of faculty members. Parsons and Platt (1973) apply structural-functional theories of society to the university and account for the American university's present structure, including the role they see the university playing in upholding core cultural and intellectual values in social life. Halsey and Trow (1971) place the study of academics in comparative perspective by offering a detailed examination of the professoriat in Britain. More recent studies have followed in the comparative and historical line of inquiry (e.g., Clark 1977, 1984, 1985, 1987b).

These and several related works in what may be called the sociology of higher education either predate the emergence of the life course perspective or are concerned, as can be inferred from the above, with matters other than the career, its organization, the subjective experience of its unfolding, and identity construction. In other words, the social psychology of career and profession has remained largely unscouted territory through much of these works. Selected points of connection with these works are, though, noted where relevant in the text.

2. Interview No. 23.
3. Ibid.
4. Ibid.
5. Ibid.
6. Ibid.
7. Ibid.
8. Ibid.
9. Interview No. 39.
10. Interview No. 25.
11. Interview No. 31.
12. Ibid.
13. Interview No. 42.
14. Interview No. 43.
15. Interview No. 58.

16. Information on mobility within the profession more generally is surprisingly difficult to come by. Lengthy searches have not found large-scale, systematic data sets that speak to the issue. Venturing with hypotheses, we would expect fairly minimal mobility in current times, given the scarcity of positions and the oversupply of labor. And we would expect this to hold as much for communitarians and pluralists as for elites: competition is extremely keen, even among the very best qualified candidates. In this respect, this sample may serve as a microworld of the larger profession.

17. The numbers of physics journals in 1973 and 1993 are compiled from the *Science Citation Index* for respective years. The Index provides the most comprehensive list of journals in physics and in other fields.

18. For related theoretic and empirical work on the meanings of age broadly applied, see Neugarten (1968, 1996) and Settersten and Mayer (1997). For related considerations of age-norms and age-related performance in science, see S. Cole (1979) and Merton and Zuckerman (1973).

19. Meyer (1986a, 1986b, 1988; Kohli and Meyer 1986) also presents an institutional perspective on the life course, that is, a perspective that stresses the high degree with which the life course for any given individual is organized by social scripts specifying not only trajectories and transitions but also how the individual is to experience and adjust to those trajectories and transitions. At any one time, self and identity become highly predictable constructs by the institutional view, because they form and function in light of structurally established terms of experience (Held [1986] offers a slightly different institutional view). "The modern system, on a number of dimensions, provides and requires a resume for each individual. Many of the elements of this resume are fixed in advance (sometimes by rules applying to everyone). Most of them are easy to anticipate, given the long chains of sequencing rules. And most of them, once established, are both fixed and important" (Meyer 1986b, 207–208). Meyer contends that measures of self and identity have been unstable because they have failed to adopt an institutional view that underscores the "rules" by which the life course organizes passage for all.

Though I too adopt a normative, institutional view of the life course, my view differs from Meyer's, which appears to place a theoretically questionable emphasis on continuity in life course passage while at the same time underplaying the role of human agency in forming life course patterns. Although my view also takes into account the expectations or "scripts" that are specified by the social systems that individuals enter (e.g., the elite social system, the pluralist social system, the communitarian social system), it additionally allows for the empirical reality that such expectations are sometimes unfulfilled and that such scripts are not always followed. As a result, self and identity in any one social system are the products not only of scripts endowed institutionally but also of the situational negotiations that compose interaction. As two metapractitioners of institutionalism in organizational analysis have aptly noted: "There has been little effort to make neoinstitutionalism's microfoundations explicit. . . . Yet any macrosociology rests on a microsociology, however tacit [or not, as the case may be]" (Powell and DiMaggio 1991, 16). My perspective, in other words, takes into account both macro and micro forces (not simply macro ones). The resulting conceptualization of the life course takes different forms in light of people's varied social participation and experience.

20. Indeed, research focused at the level of the individual has found that differences in measured ability do not predict levels of performance (Bayer and Folger 1966; Cole and Cole 1973, 69).

21. For example, many universities have begun instituting various kinds of rewards to recognize performance. Endowed professorships are one such example. Endowed professorships originated in elite universities, but other institutions have taken up the practice. Riesman (1956) characterizes universities as being in

a "snake-like progression," in which they constantly imitate or try to catch up with the set of institutions directly above them. The "head" of the snake—the elite—sets the tone, defining the standards and interests; other parts of the snake feel the tone and attempt to follow in successive reverberation. Similarly, Janowitz (1960) invokes a biology metaphor: the "elite nucleus" of a profession controls all other parts of the whole by setting standards and making directives that others attempt to follow. Most important, endowed professorships operate as a mechanism of social control, regulating performance with the hope that the crown-bearers—and their institutions—may themselves be elite. The proliferation of endowed professorships is much like that of credit cards: not only can you have a card, but you can become a gold member, even a platinum member. Professors are likewise "distinguished," "more distinguished," and "even more distinguished." So many categories of the elite have been created that one loses track of who's who and exactly what each category means. Nevertheless, endowed professorships stand as a source of esteem.

22. My use of the term "boundary work" is distinct from Gieryn's (1983). Gieryn uses the term to describe an ideology found among scientists in their attempts to present a favorable public image by contrasting their work with nonscientific intellectual or technical activities. In his case, the term is used to account for the cultural authority and credibility of modern science. Along similar lines, Abbott (1988) discusses boundary work in terms of jurisdiction—the claims that groups make to control certain bodies of knowledge and practices, as in the dispute between the clergy and psychiatry over control of the "personal problems" jurisdiction. In my case, "boundary work" is used to account for how individuals and groups create identities, as in rhetorical ploys that highlight similarities and differences between groups.

23. Interview No. 45.

24. Interview No. 46.

25. Interview No. 43.

26. Interview No. 38.

27. Interview No. 42.

28. Interview No. 6.

29. Interview No. 43.

30. Interview No. 44.

31. Interview No. 53.

32. Parallels are found in Lortie (1975, esp. chapter 4), who, in a study of primary and secondary schoolteachers, accounts for the relative lack of staging in careers centered on teaching. He notes, in ways as germane to this study as to his own: "Staged careers produce cycles of effort, attainment, and renewed ambition. In tying the individual to the occupation they give him a stake in its future. Staging gives reality and force to the idea of the future; it generates effort, ambition, and identification with the occupation" (Lortie 1975, 85). Lortie thus hypothesizes that the relative lack of stages in the teaching career leads to the predominance of a present, versus future, orientation to work.

Chapter 3

1. There are several related studies of professional socialization, and I confine myself to citing those that have been especially helpful in my thinking. See Becker

et al., *Boys in White* (1961); Becker and Carper, "The Development of Identification with an Occupation" (1956a) and "The Elements of Identification with an Occupation" (1956b); Becker and Strauss, "Careers, Personality, and Adult Socialization" (1956); Cookson and Persell, *Preparing for Power* (1985); Faulkner, "Coming of Age in Organizations" (1974), *Hollywood Studio Musicians* (1985), and *Music on Demand* (1987); Habenstein, *The Career of the Funeral Director* (1954); Hughes, *Men and Their Work* (1958); Janowitz, *The Professional Soldier* (1960); Lortie, "From Laymen to Lawmen" (1959); Merton, Reader, and Kendall, *The Student-Physician* (1957); Miller, *Prescription for Leadership* (1970); Osherson, *Holding On or Letting Go* (1980); Reimer, "Becoming a Journeyman Electrician" (1977); Shaw, *The Jack-Roller* (1930); Sofer, *Men at Mid-Career* (1970); Weiss, *Staying the Course* (1990); Zussman, *Mechanics of the Middle Class* (1985).

Boston's public broadcast station, WGBH, also has contributed to the study of professional socialization through the production of *The Making of a Doctor* (WGBH 1995), a major sociologically sensitive study that followed a group of Harvard medical students through their training and into the early phases of their careers. Though this piece merits attention in its own right, I have also found it useful—as may others—as a component of undergraduate courses.

For a fictional (and somewhat saccharine but nevertheless intriguing) account of how the lives of five extraordinary members of the Harvard class of 1958 unfold, see Segal's *New York Times* bestseller, *The Class* (1985).

2. Interview No. 16.

3. Interview No. 19.

4. For a discussion of the "ethos" of science, including the four norms that are said to undergird it—universalism, communism, disinterestedness, and organized skepticism—see Merton's foundational essay "The Normative Structure of Science" (1973b), first published in 1942.

5. For further discussion on the importance of recognition to science careers, see Merton (1973a, 1973d); Zuckerman (1989); Gaston (1973, esp. chapters 4 & 5); and Hagstrom (1965, esp. chapter 2). Gustin (1973) takes issue with viewing recognition as an adequate key to the motivation of scientists. He asserts that a significant fraction of the scientific community publishes very little. But recognition need not be thought of so parochially. Recognition takes diverse forms, with only a part, albeit large, stemming from research productivity. As Zuckerman states, "Teaching, administration, and citizenship in the community of science also contribute to the advancement of knowledge in less obvious ways" (Zuckerman 1988, 528).

6. A list of the top-ten departments of physics in the United States can be compiled from the assessment performed by the National Research Council (NRC) in Goldberger, Maher, and Flattau (1995). This ranking roughly corresponds to the better-known and popular, but widely criticized, *U.S. News & World Report* ranking (see, for example, *U.S. News & World Report* [1993]), which is produced periodically and to which many of the scientists here refer in their narrative accounts. Because of ties in scoring, both of the rankings include eleven schools. The NRC ranking includes, from highest to lowest: Harvard, Princeton, M.I.T., Berkeley, Cal Tech, Cornell, Chicago, Illinois, Stanford, University of California–Santa Barbara, and Texas. The *U.S. News & World Report* ranking includes, from highest

to lowest: Cal Tech, Harvard, M.I.T., Berkeley, Stanford, Princeton, Cornell, Chicago, Illinois, Yale, and Columbia.

7. Interview No. 42.

8. Richard P. Feynman (1918–88), awarded the Nobel Prize in Physics in 1965.

9. Hans A. Bethe (1906–), awarded the Nobel Prize in Physics in 1967.

10. Interview No. 20.

11. Murray Gell-Mann (1929–), awarded the Nobel Prize in Physics in 1969.

12. Interview No. 55.

13. Interview No. 12.

14. Interview No. 56.

15. Schwartz's (1981) major book attempts to explain human beings' cross-cultural propensity to use spatial metaphors—such as high and low, up and down, above and below—to represent social inequalities in various arenas. Cognitive and moral dimensions are linked so that what is above is usually said to be superior and what is below is said to be inferior. An example can easily be found in the manner in which educational institutions are popularly classified. Gradations of "quality" are associated with a ranking on a vertical scale, and estimations of individual ability and achievement tend to be imputed from the prestige of the organization in which one is a member. Schwartz claims that a vertical code is rooted in the personal experience of childhood and is transferred to the adult world. Vertical classification becomes a social code that is used as a universal language with both cognitive and moral meanings. Whereas systems of vertical classification serve as a way of organizing complex social forms, they often differentiate on a gross level only (e.g., better, lesser) and reify what may often be erroneous distinctions in the moral worth of the very objects being classified.

16. Interview No. 22.

17. Interview No. 28.

18. Interview No. 40.

19. Interview No. 54.

20. Ibid.

21. Steven Weinberg (1933–), awarded the Nobel Prize in Physics in 1979.

22. Interview No. 12.

23. Enrico Fermi (1901–54), awarded the Nobel Prize in Physics in 1938.

24. Interview No. 1.

25. Interview No. 11.

26. True equity and views of equity are of course distinct matters.

27. A reward system defined as *inequitable* leads the Coles to conclude that a rationale is provided for violating norms: engaging in deviant behavior in science such as using illegitimate means to gain recognition (Cole and Cole 1973, 255–261). That appears to be a leap in logic. Even if people view their situation as unfair, the propensity to engage in deviant behavior works as a function of evaluating costs and benefits. The chances of achieving substantial recognition through fraudulent means is, first of all, slim, but even then, the penalties for being caught likely far outweigh (at least for most people) the pleasures that might come from fraudulently achieved recognition. Some people may view "the system" as unjust in certain ways, but they continue to work within it because there are no plausible alternatives. In fact, many people develop, as further accounts

will demonstrate, a "hardened identity" that often reflects individual failings as much as it does systemic ones.

28. Glaser (1964b) alternately speaks of "cooling-down" rather than "cooling-out" to underscore the comparative, as opposed to the absolute, conception of failure: though "cooled," scientists normally do not lose all commitment to their profession but readjust and redirect it. In principle this is consistent with Goffman's original usage (the mark comes to be satisfied with some alternative), and so I use the original term.

29. None of the communitarians earned their doctoral degrees at communitarian institutions. Only four of nineteen completed their undergraduate study at a communitarian school.

30. Interview No. 31.

31. Ibid.

32. Interview No. 25.

33. Interview No. 27.

34. Interview No. 40.

35. Interview No. 26.

36. Interview No. 27.

37. Interview No. 32.

38. Interview No. 27.

39. Interview No. 30.

40. On the glut of academic labor in physics, see Kirby and Czujko (1993) and Ellis (1993). On novel solutions to the problem, see Taubes (1994). See also the report from a National Research Council panel convened to assess possible reforms in the graduate education of scientists, including reforms that would expand career opportunities (Committee on Science, Engineering, and Public Policy 1995).

41. Interview No. 34.

42. Interview No. 39.

43. Interview No. 55.

44. Interview No. 59.

45. Interview No. 56.

46. Interview No. 50.

47. Interview No. 46.

48. Interview No. 50.

49. Interview No. 49.

50. Interview No. 23.

51. Interview No. 6.

52. Interview No. 14.

53. Interview No. 16.

54. Interview No. 13.

55. On the surface, it might seem contradictory to call worlds "communitarian" when individual communitarians seem the most at odds with the world in which they work. Recall, however, the basis on which this group is identified. Communitarianism expresses the key *collective aspiration* of members who belong to this world. Their grounds for choice and action stem first and foremost from local ties to the institution itself. Although these scientists are communitar-

ian in their aspirations, not all of them, as we have seen, are necessarily communitarian in their happiness.

56. Here, as elsewhere, the names I use are pseudonyms. I have attached names to accounts that I cite at length in order to contextualize the narrator and to help keep the accounts straight.

57. Interview No. 20. This interview is also the source for the following quotations from the "Feldman" account.

58. This genre in and of itself represents a distinction among worlds of science. If elites write textbooks (work that many of them are disinclined to do because it does not represent original work), they are most likely to do so in late career or even at the end of the career, when opportunity costs are fewer. Pluralists and communitarians are less likely to abide by the same norm. In these worlds textbook writing is more often seen at earlier career stages, in part because pluralist and communitarian worlds are in general more allied with the teaching functions that textbook writing serves.

59. Interview No. 3.

60. Some scientists joked with me about how the Nobel Prize can essentially kill a career rather than boost it. See Zuckerman (1977) on various matters related to the careers of Nobelists, including the consequences of renown on productivity.

61. Interview No. 16.

62. Interview No. 51.

63. Interview No. 57.

64. Interview No. 58.

65. In the subsample of elites, I interviewed two former chairs of departments of physics and a director of a research center. The chairs had served three-year terms, highlighting a common elite pattern in which people normally serve in fixed-term, rotating appointments, although not all members of a department rotate into administrative positions. The research director also was serving in a fixed-term capacity (five years, with a possibility of renewal). These characteristics underscore the low tolerance that elites generally have for career administrators.

In the subsample of pluralists, no past or present administrators were found, but in general, people who enter high-level administrative positions (e.g., deanships, provostships, and presidencies) in this world are likely to be "excused" from research and/or teaching duties, whereas deans and some provosts in the elite often remain active on the research or scholarly front. These differences may stem from structural arrangements across types of academic worlds. Pluralist worlds, usually made up of large public state institutions, conceivably place considerably more time constraints on administrators, who may be held accountable for larger academic units and a greater number of personnel and policy decisions. Professionally, pluralist administrators also may take an active role in the science policy matters that surface on the national scene, becoming in effect diplomats, guardians, boosters, or policy pioneers. Although pursued in a different vein, their influence on science may in some cases equal or far exceed that of scientists who follow a more traditional research path.

In the subsample of communitarians, I interviewed three former chairs who had served in three-year, rotating terms; one of them served one term, another one and a half terms, and the third one term as chair and also one year as an assistant dean, two years as an assistant vice-chancellor, and six months as an acting dean. One other communitarian followed an administrative path that was considerably more lengthy, and I take up his case below.

66. Interview No. 28.

Chapter 4

1. After C. N. Yang (1922–) and Robert L. Mills (1927–).
2. After Subrahmanyan Chandrasekhar (1910–95).
3. After Brian David Josephson (1940–).
4. After Lewis Fry Richardson (1881–1953).
5. After Sir George Paget Thomson (1892–1975).
6. After Hans Wilhelm Geiger (1882–1945).
7. After Michael Faraday (1791–1867).
8. After Peter Joseph Debye (1884–1966) and Erich Armand Arthur Joseph Hückel (1896–1980).
9. After Sir Charles Wheatstone (1802–75).
10. After Ernst Heinrich Weber (1795–1878) and Gustav Theodor Fechner (1801–87).
11. Interview No. 23.
12. Ibid.
13. Ibid.
14. Ibid.
15. Interview No. 2.
16. Interview No. 5.
17. Ibid.
18. Interview No. 8.
19. Ibid.
20. Ibid.
21. Interview No. 10.
22. Interview No. 14.
23. Interview No. 13.
24. Ibid.
25. As I explained earlier, "failure" should not be construed in absolute terms, although it is conceivable that a tiny fraction of scientists might consider themselves absolute failures as a result of the most peculiar circumstances. Used here, "failure" is another way of stating relative success.
26. Interview No. 33.
27. Interview No. 36.
28. Interview No. 31.
29. A report by the National Research Council (National Research Council 1991) on the consequences for universities of ending mandatory retirement for professors found that the faculty members most likely to remain in the professoriat on a full-time basis after age sixty-five and also after age seventy were those at a small handful of elite research universities. By and large, the effects of uncapping mandatory retirement were predicted to be minimal for the vast majority of institutions outside of this category.

30. For an interesting social history of retirement, see Graebner (1980).
31. Interview No. 35.
32. Interview No. 34.
33. Interview No. 30.
34. Interview No. 25.
35. Interview No. 52.
36. Ibid.
37. Interview No. 59.
38. Ibid.
39. Interview No. 12.
40. Interview No. 3.
41. Interview No. 1.
42. Interview No. 7.
43. Interview No. 17.
44. Interview No. 57.
45. Ibid.
46. Ibid.
47. Interview No. 58.
48. Interview No. 46.
49. Interview No. 38.
50. Interview No. 42.
51. Interview No. 26.
52. Interview No. 29.
53. Interview No. 42.
54. Interview No. 39.
55. Interview No. 56.
56. Ibid.
57. Interview No. 23.

58. Haber (1991) has offered a cultural history of the professions by explaining how they organized over three distinct periods in response to class distinctions. In an excellent work, Bledstein (1976) examines a variation of the historical theme. He traces the rise of "professional culture" in modern life and assesses its development in terms of the consequences for higher education.

59. It would indeed be of interest and relevant to policy concerns to examine which professions are most competitive, in the sense of entering and moving within them. This is, of course, an empirical question. To speculate, careers in science currently bear more similarities to professional sports than to professions such as medicine or law. Science and sports are comparatively difficult to enter, and very few individuals "make it to the majors." By contrast, medicine and law, though also difficult to enter compared with most other occupations, offer individuals a range of institutions from which to get their training, and in many instances (especially outside of academic medicine and law) these institutions do not effect place of future employment. The standardized training of doctors and lawyers also would appear to facilitate greater mobility in the professions. In the wide worlds of medicine and law, practitioners may be, in general, less concerned about where people "went to school" than are members of the academy. Doctors and lawyers who do show such concerns likely compose the elite of their fields,

where such comparisons are a sacred element of occupational culture. It is unclear what other lines of work are as structured as is science by the place of first employment.

60. This of course assumes that we hold individual ability constant for the sake of the exercise. Other issues come into play. Among them are cognitive developments in a field. Older physicists sometimes remarked that it was easier to make major discoveries in earlier times because the field was less mature. Currently, physics can be described as being in a mature state, to the extent that some have wondered about its vitality and future promise (Lindley 1993). Gains in knowledge within physics proper would appear to be smaller today than they were fifty years ago. As one of the physicists commented:

> To some extent, physicists have been spoiled because the first half of this century has been so revolutionary—there was quantum mechanics and relativity and things like that. And even in the seventies, when I was a graduate student, it was the heyday of quantum mechanics in terms of new experimental results. That was really exciting. Now, of course, there are a lot of exciting things, but each step is so [big]; to make progress is so costly in manpower and funding. (Interview No. 52)

How does growth in a field occur once the field comes close to stagnation? By one view (Abbott 1988), a field reorganizes along different cognitive boundaries that represent innovative (or politically more powerful and persuasive) jurisdictions of expertise. Following this view, we may one day no longer see departments of physics in universities but rather departments of something else that house new cognitive alignments. Hankins (1985) has provided an insightful historical view of how science in the eighteenth century was organized into fields different from those familiar to us today, a change that reflects the cognitive status of esoteric knowledge at different times. In a collective biography of American astronomers up to 1940, Lankford (1997) addresses the rise of astronomy in the United States, together with the blossoming of careers in that field. The evolving process of professional growth and decline characterizes many developments in the fields of education (Clifford and Guthrie 1988), biology (Roush 1997), classics, and library science, among others—fields that increasingly see their jurisdictional claims lost to competing parties.

61. Predictions of a renaissance in faculty hiring in the 1990s, due largely to anticipation of large-scale retirement (Bowen and Sosa 1989), appear to have been somewhat off target.

62. For comparative purposes, according to the U.S. Bureau of Labor Statistics, the unemployment rate for the entire U.S. population in 1993 was 6.8 percent (National Science Board 1996, 3–5).

63. The essays in Galison and Hevly (1992) provide a historical account of the rise of large-scale scientific research in the twentieth century.

64. Interview No. 1.

65. Interview No. 7.

66. Interview No. 2.

67. People in less austere times may well have been competitive. Was Newton? Einstein? Perhaps, but the point is that now, institutionalized stakes are higher, and competition has become—for most scientists—an occupational prerequisite.

Informally, I recall a senior colleague asking earnestly why graduate students had grown ruthlessly competitive. He felt that they had been far more cooperative in earlier times, freely helping one another in their work. What seemed lost in the question was the observation of competition among the department members themselves. Students and their teachers may compete for different things, but the things for which they each compete are increasingly scarce. There is also a "like parent, like child" phenomenon: students learn how to act by embracing many of the qualities manifest in their teachers. Whether some behavioral patterns serve communities and their members well (i.e., whether they are functionally useful) is a question whose answer is best sought elsewhere.

68. Austere socioeconomic conditions raise performance expectations. This assertion is premised on conditions of the academic labor market and other infrastructural factors, such as availability of funding. The reverse likely does not hold. That is, when market and fiscal conditions loosen, performance expectations do not lower but likely stay the same because a precedent has been achieved. The lowering of standards would result in a loss of prestige and recognition, which would run counter to the institutionalized goals of science. (This hypothesis has not been tested, however, because in general conditions in education since the early 1970s have remained austere.) Thus, institutionalized norms governing performance likely maintain a standard even when external conditions decompress.

69. Scientists characteristically receive tenure after four to five years of being an assistant professor; in many other fields, the length of time is characteristically six to seven years.

70. The extent of institutionalization of the postdoc varies among academic fields. Widespread in the natural sciences, institutionalization of the postdoc is less common among the social sciences and rarer still among the humanities, but this appears to be changing moderately over time. In the social sciences, the postdoc often takes root in fields and specialties that bear the most similarities to the natural sciences (i.e., those that tend to be the most quantitative), such as economics, and in the area of demography, most often housed within sociology.

71. This is, of course, oversimplified. Characteristics of the graduate school stage—productivity, mentor, institutional prestige—also affect the later career. But the point remains that even with those factors taken into account, the postdoctoral stage is comparatively the most crucial for subsequent professional prosperity.

72. Interview No. 5.

73. "Period effects" should not, of course, be used as an ambiguous umbrella term to explain the impact of macro events on individuals. More precisely, a period effect normally encompasses a range of social forces and processes that impinge on individual lives, often in different ways (Elder 1974; Elder and O'Rand 1995).

74. Interview No. 19.

75. Interview No. 32.

Chapter 5

1. Thanks go to Harriet Morgan for calling this passage to my attention.

2. Interview No. 12.

3. Interview No. 23.

4. Interview No. 1.

5. Interview No. 16.

6. Interview No. 14.

7. Interview No. 4.

8. Among elites' other peculiarities was a concern with privacy in salary matters. Of the sixty scientists, seven declined to furnish salary information. Of these, four were elites and were especially prominent within the science community, occupying what Zuckerman (1977) has called the "ultra elite." This situation can be read in two ways. First, celebrities are especially guarded about such information. The release of salary can jeopardize reputations, especially for those who play to wide public audiences. This fear is evident even with severe assurances of confidentiality. As one of my respondents shot back, "Nothing is confidential!" In addition, salary is considered by many to be an objective measure of individual worth. Elites may be equally embarrassed whether their salaries are comparatively low or outrageously inflated. Second, elite universities in particular appear to have grown sensitive to the release of individual, and even aggregate, salary information. (Public institutions are often legally compelled to furnish salary information.) This is perhaps a consequence of the vigor with which elite universities compete for candidates. It is also a means to protect their prestige. Among other results, the practice reinforces the air of seriousness, mystifying while keeping sacred the academic role that people perform.

9. Interview No. 7.

10. Interview No. 1.

11. Ibid.

12. Interview No. 23.

13. Ibid.

14. Interview No. 25.

15. Ibid. This scientist's department each year formally evaluates its faculty members on standard criteria of research (including grantsmanship and conference presentations), teaching, and service. A scientist's performance is scored under each area, and the faculty members are grouped according to a five-tier scale: A, B, C, D, E. A composite score can be derived, resulting in a ranking of all members. In an annual meeting with the department chair, individuals are informed of what tier they placed in the various areas and their overall ranking. The process itself is overseen by a committee of four faculty members, whose terms rotate on a multiyear cycle. At least in the case of this scientist, there is apparently no irony in or contradiction between being ranked "number one" in this type of department and adopting the style of life the respondent so nonchalantly describes.

16. Interview No. 27.

17. Interview No. 31.

18. Interview No. 40.

19. Interview No. 49.

20. Interview No. 51.

21. Interview No. 53.

22. Interview No. 46.

23. Interview No. 56.

24. This principle may of course operate among elites, but elites have more to fall back on professionally when loftier ambitions elude them.

25. Interview No. 46.

26. For example, the biological sciences have more women practitioners compared with other fields and have proven to be more hospitable in terms of career advancement than have other fields (Sonnert and Holton 1995). In general, women seem to encounter the fewest career obstacles in fields where they are represented in relatively high numbers. Sonnert and Holton (1995, 51) suggest that a concentration of a group, as in the biological sciences, creates a "critical mass" that inactivates gender stratification and segregation.

27. Readers who are in search of an extensive examination of gender and science are directed to several excellent sources, including Sonnert and Holton (1995), who offer the most recent and comprehensive assessment, as well as J. Cole (1979), Long and Fox (1995), and Zuckerman, Cole, and Bruer (1991). For a rich history of women scientists in America, see Rossiter (1982, 1995).

28. The senior woman of the elite held a research position at a laboratory for seven years before taking an academic post. The laboratory is affiliated with the institution where she took the post and where she has worked ever since.

29. Female scientists are less likely to be married or to have children than are male scientists and women in the general population (Astin 1969; Centra 1974; Long 1990; Long and Fox 1995).

30. Domestic labor roles refer not only to child care and child-rearing but also, and more generally, to household maintenance, which may or may not involve children.

31. Sonnert and Holton (1995) do not supply evidence to support the claim that women drop out of science more often than men. The authors do find, however, that on average women spend more time on the doctorate than men, and they attribute this difference to parenthood demands and to women's greater propensity to interrupt studies or to study part-time. Interruptions in the educational career and parenthood demands are of course related in principle.

32. In general, discussion of the women scientists requires extra care to ensure anonymity. Because they are so few in number, providing information that would inform an understanding of their careers and that would be of general interest to readers is difficult and often not feasible.

33. Her husband is also a distinguished academic.

34. Interview No. 14.

35. Interview No. 40.

36. Interview No. 14.

37. Interview No. 9.

38. In the most recent and comprehensive study of gender and science to date, the following conclusion was drawn based on information collected from a national sample of scientists across many fields: "On the whole, the women in our sample did not do extremely worse than the men; very large and very obvious gender differences and disparities were absent" (Sonnert and Holton 1995, 164).

39. Research has consistently shown that on average, women scientists tend to be less productive in publication than men matched for age, doctoral institution, and field (J. Cole 1979; Cole and Zuckerman 1984, 1987; Reskin 1978; Sonnert and Holton 1995). But the research has also found that marriage and family obliga-

tions do not account for this difference (Cole and Zuckerman 1987; Hargens, Mc-Cann, and Reskin 1978). Even for professional rank, marriage has been shown to have a positive effect, rather than a negative one, and the effects of having children have been found to be weak or nonexistent (Long and Fox 1995). Collaboration practices also have not explained productivity disparities between men and women (Cole and Zuckerman 1984).

Researchers appear to agree that gender differences in research productivity remain puzzling (Cole and Zuckerman 1984; Zuckerman, Cole, and Bruer 1991). Some of the most recent and suggestive insights, however, point not to marriage, motherhood, collaboration, or related factors to explain the difference but to effects stemming from the low *density* of women in science. Sonnert and Holton (1995) suggest that the lack of high numbers of women in science leads to a distinctive "scientific style": women frequently adopt a more meticulous and perfectionist approach to research and tend to more strenuously uphold traditional standards of science, such as carefulness, replicability, and connection to fundamental ideas. Perhaps women embrace this style because their marginal status compels them to adopt extra-high standards of conformity in order to be considered legitimate members of the scientific community (see Hermanowicz 1996). Sonnert and Holton seize on the notion of the accumulation of advantages and disadvantages to explain the process of gender inequality: initially small differences are accentuated in later career stages, resulting in disparities of performance.

40. Interview No. 42.

41. Interview No. 52.

42. This is not to suggest that women (or men) fail to bring a "seriousness of purpose" to their family roles, although this may of course be true in specific cases. Women and men may embark on family "careers" with great intentions, even with a competitive spirit, but "ambition" is not the adjective historically used to describe their behavior.

43. Recall that several men of the pluralist and communitarian worlds exhibit elite-like characteristics, but not *all* of the men.

44. Interview No. 14.

45. Ibid.

46. Interview No. 9.

47. Interview No. 52.

48. Interview No. 42.

Chapter 6

1. My use of the term "master narrative" differs from and should not be confused with its use by postmodernists, who refer to master narratives for much different reasons (i.e., as a critique of history). My use is drawn indirectly from the sociologist Everett C. Hughes (see the introduction), who developed the idea of "master status" (1958)—a dominant social position that guides an understanding of all other characteristics of the person holding the position.

2. Furthermore, when I say "dominant," I mean dominant. The term is used in the strict empirical sense: a *majority* of cases (in any given world) fall into the respective central pattern.

3. It is for these reasons as well that I have argued that "social world" is a better

way to conceptualize professions than is "strata," which major segments of the sociology of science literature adopt to explain contextual distinctions (see the introduction). Although contexts are distinct in ways, they are similar in others. By speaking of social "worlds" of science rather than clearly delineated "strata" of science, we are better equipped to ascertain and investigate similarity and difference.

4. Just as elite worlds sometimes include pluralists and some communitarians (and vice versa), industrial societies include farming, and agrarian societies often include, or come to include, industry. Yet in each case, a collective identity arises that meaningfully locates groups—and their key tendencies—within an array of types.

5. See the section "Feeling the Effects of Time" in chapter 4.

6. This is a hypothesized relationship confirmed by other studies (see Fox 1983; and Reskin 1979). Information on mentors for *every* member of this specific sample was not available.

7. Of these factors, early success is the most closely linked to doctoral origins. Successful young scientists are most likely to have been trained at elite schools. Divorce is also most common among elites (see chapter 5), though only marginally more frequent than among communitarians, who are more often trained at more varied institutions. This is not to suggest that all people who experience these factors end up taking an alternate professional course; we can conclude only that they are more prone to. Additionally, there are undoubtedly people who have such experiences but who continue to conform to the moral career. In this, we are led to Clausen (1993), who suggests that those who are "planfully competent" beginning in their youth not only are able to attain high occupational statuses but also are in a better position to weather, and advance beyond, the emotional storms of life.

8. In a highly suggestive work, Bruner (1996) discusses narrative in terms of a concern with education and the broader advancement of society. Narrative imparts knowledge, including knowledge of the culture in which one attempts to live competently. On the one hand, access to different narratives establishes a basis of social differentiation and stratification. On the other hand, according to Bruner, the extension of rich narratives to wide groups—principally through schooling—is a prime means of individual enlightenment and societal advancement. Such a view necessarily renews concern with the selection, training, and socialization of prospective teachers and with the content of what they transmit to the young.

9. For example, it was difficult (though not impossible) for scientists to lie about or overstate their achievements because their curricula vitae documented many of the objective features of their careers. More often, embellishment is found in lives not yet lived: in the future self. Compare the role of embellishment in science careers with its role in careers of the homeless (Snow and Anderson 1993, esp. chapter 7). Scientists are at greater risk of looking like fools because they are typically located within a network of people who have access to information about the objective facts of their records and personal histories. By contrast, embellishment among the homeless becomes an integral component of identity construction. Few or no other people are in a position to dispute status claims. Some claims may be valid while others may not be, and an audience, though likely

skeptical of stories told by those of marginal status, lacks any firm basis on which to repudiate the claims. Aside from enhanced images of the future self, embellishment by scientists usually takes more subtle forms. Notable examples are those that highlight affiliation as marks of status: "Before joining the faculty at Berkeley, I spent my early career at a university on the banks of the Charles River that shall remain nameless," or "When I was a graduate student, Feynman used to tell us . . ."

10. Young academics may belittle, forget about, or never learn of the accomplishments of their seniors.

11. There are, of course, exceptions, such as Sewall Wright (the noted zoologist), who after retirement from the University of Chicago went on to have a "second," multidecade career at the University of Wisconsin.

12. Interview No. 21.

13. Interview No. 6.

14. Interview No. 23.

15. Tom Russell, "Box of Visions," Rounder Records (1994).

16. William Butler Yeats, "To a Friend Whose Work Has Come to Nothing."

REFERENCES

Abbott, Andrew. 1988. *The System of Professions: An Essay on the Division of Expert Labor.* Chicago: University of Chicago Press.

Allison, Paul D., J. Scott Long, and Tad K. Krauze. 1982. "Cumulative Advantage and Inequality in Science." *American Sociological Review* 47:615–625.

Allison, Paul D., and John A. Stewart. 1974. "Productivity Differences among Scientists: Evidence for Accumulative Advantage." *American Sociological Review* 39:596–606.

American Council on Education. 1992. *American Universities and Colleges.* New York: Walter de Gruyter.

American Institute of Physics. 1995. "1994 Initial Employment Follow-up of 1993 Physics Degree Recipients." *AIP Report: Education and Employment Statistics Division.* No. R-282.17 (July).

———. 1994. "1993–94 Academic Workforce Report." *AIP Report: Education and Employment Statistics Division.* No. R-392.1 (December).

———. 1993. *Graduate Programs in Physics, Astronomy, and Related Fields, 1993–94.* New York: American Institute of Physics.

Anderson, Benedict. 1983. *Imagined Communities: Reflections on the Origin and Spread of Nationalism.* New York: Verso.

Aristotle. [1986]. *Nicomachean Ethics.* Translated by Martin Ostwald. New York: Macmillan.

Arthur, Michael B., and Barbara S. Lawrence. 1984. "Perspectives on Environment and Career: An Introduction." *Journal of Occupational Behaviour* 5:1–8.

Astin, Helen S. 1969. *The Woman Doctorate in America: Origins, Career, and Family.* New York: Russell Sage Foundation.

Augustine, Saint. [1960]. *The Confessions of St. Augustine.* Translated by John K. Ryan. New York: Doubleday.

Bailyn, Lotte. 1989. "Understanding Individual Experience at Work: Comments on the Theory and Practice of Careers." In Michael B. Arthur, Douglas T. Hall, and Barbara S. Lawrence (eds.), *Handbook of Career Theory*, 477–489. Cambridge: Cambridge University Press.

Baltes, Paul B. 1979. "Life-Span Developmental Psychology: Some Converging Observations on History and Theory." In Paul B. Baltes and Orville G. Brim, Jr.

(eds.), *Life-Span Development and Behavior,* vol. 2, 256–281. New York: Academic Press.

Baltes, Paul B., and John R. Nesselroade. 1984. "Paradigm Lost and Paradigm Regained: Critique of Dannefer's Portrayal of Life-Span Developmental Psychology." *American Sociological Review* 49:841–847.

Barber, Bernard. 1952. *Science and the Social Order.* Glencoe, Ill.: Free Press.

Barron's Educational Series. 1996. *Profiles of American Colleges.* Hauppauge, N.Y.: Barron's Educational Series.

Bayer, A. E., and J. F. Folger. 1966. "Some Correlates of a Citation Measure of Productivity in Science." *Sociology of Education* 39:381–389.

Becker, Howard S. 1982. *Art Worlds.* Berkeley: University of California Press.

———. 1960. "Notes on the Concept of Commitment." *American Journal of Sociology* 66:32–40.

Becker, Howard S., and James Carper. 1956a. "The Development of Identification with an Occupation." *American Journal of Sociology* 61:289–298.

———. 1956b. "The Elements of Identification with an Occupation." *American Sociological Review* 21:341–348.

Becker, Howard S., Blanche Geer, Everett C. Hughes, and Anselm L. Strauss. 1961. *Boys in White: Student Culture in Medical School.* Chicago: University of Chicago Press.

Becker, Howard S., and Anselm L. Strauss. 1956. "Careers, Personality, and Adult Socialization." *American Journal of Sociology* 62:253–263.

Ben-David, Joseph. 1991. *Scientific Growth: Essays on the Social Organization and Ethos of Science.* Edited by Gad Freudenthal. Berkeley: University of California Press.

Berger, Bennett M. 1990. "Looking for the Interstices." In Bennett M. Berger (ed.), *Authors of Their Own Lives: Intellectual Autobiographies by Twenty American Sociologists,* 152–164. Berkeley: University of California Press.

Berger, Peter L., and Thomas Luckmann. 1966. *The Social Construction of Reality: A Treatise in the Sociology of Knowledge.* Garden City, N.Y.: Doubleday.

Bertaux, Daniel, and Martin Kohli. 1984. "The Life Story Approach: A Continental View." *Annual Review of Sociology* 10:215–237.

Binstock, Robert H., and Linda K. George (eds.). 1990. *Handbook of Aging and the Social Sciences.* 3d ed. San Diego: Academic Press.

Bissinger, H. G. 1990. *Friday Night Lights: A Town, A Team, and a Dream.* New York: Harper Collins.

Blau, Peter M. 1973. *The Organization of Academic Work.* New York: John Wiley.

Bledstein, Burton J. 1976. *The Culture of Professionalism: The Middle Class and the Development of Higher Education in America.* New York: W. W. Norton.

Bottomore, Tom. 1964. *Elites and Society.* London: C. A. Watts.

Bowen, William G., and Neil L. Rudenstine. 1992. *In Pursuit of the Ph.D.* Princeton: Princeton University Press.

Bowen, William G., and Julie Ann Sosa. 1989. *Prospects for Faculty in the Arts and Sciences: A Study of Factors Affecting Demand and Supply, 1987 to 2012.* Princeton: Princeton University Press.

Brim, Gilbert. 1992. *Ambition: How We Manage Success and Failure throughout Our Lives.* New York: Basic.

Brim, Orville G., Jr., and Jerome Kagan. 1980. "Constancy and Change: A View of the Issues." In Orville G. Brim, Jr., and Jerome Kagan (eds.), *Constancy and Change in Human Development*, 1–25. Cambridge: Harvard University Press.

Brim, Orville G., Jr., and Carol D. Ryff. 1980. "On the Properties of Life Events." In Paul B. Baltes and Orville G. Brim, Jr. (eds.), *Life-Span Development and Behavior*, vol. 3, 367–388. New York: Academic Press.

Brines, Julie. 1994. "Economic Dependency, Gender, and the Division of Labor at Home." *American Journal of Sociology* 100:652–688.

Brint, Steven. 1994. *In an Age of Experts: The Changing Role of Professionals in Politics and Public Life*. Princeton: Princeton University Press.

Bronfenbrenner, Urie. 1979. *The Ecology of Human Development*. Cambridge: Harvard University Press.

Brooks, Peter. 1984. *Reading for the Plot: Design and Intention in Narrative*. Cambridge: Harvard University Press.

Bruner, Jerome. 1996. *The Culture of Education*. Cambridge: Harvard University Press.

———. 1990. *Acts of Meaning*. Cambridge: Harvard University Press.

Buchmann, Marlis. 1989. *The Script of Life in Modern Society*. Chicago: University of Chicago Press.

Bulmer, Martin. 1984. *The Chicago School of Sociology: Institutionalization, Diversity, and the Rise of Sociological Research*. Chicago: University of Chicago Press.

Cahill, Spencer, Gary Alan Fine, and Linda Grant. 1995. Dimensions of Qualitative Research. Chap. 23 in Karen S. Cook, Gary Alan Fine, and James S. House (eds.), *Sociological Perspectives on Social Psychology*, 605–629. Boston: Allyn and Bacon.

Caplow, Theodore, and Reece J. McGee. 1958. *The Academic Marketplace*. New York: Basic.

Carlyle, Thomas. [1904] 1966. *On Heroes, Hero-Worship, and the Heroic in History*. Lincoln: University of Nebraska Press.

Carnegie Commission on Higher Education. 1994. *A Classification of Institutions of Higher Education*. Princeton: Carnegie Foundation for the Advancement of Teaching.

———. 1973. *A Classification of Institutions of Higher Education*. Princeton: Carnegie Foundation for the Advancement of Teaching.

Caspi, Avshalom, and Daryl J. Bem. 1990. "Personality Continuity and Change across the Life Course." In L. A. Pervin (ed.), *Handbook of Personality*, 549–575. New York: Guildord.

Caspi, Avshalom, Glen H. Elder, Jr., and Ellen S. Herbener. 1990. "Childhood Personality and the Prediction of Life-Course Patterns." In Lee Robbins and Michale Rutter (eds.), *Straight and Devious Pathways from Childhood to Adulthood*, 13–35. Cambridge: Cambridge University Press.

Centra, J. D. 1974. *Women, Men, and the Doctorate*. Princeton: Educational Testing Service.

Chandrasekhar, S. 1987. *Truth and Beauty: Aesthetics and Motivations in Science*. Chicago: University of Chicago Press.

Chinoy, Ely. 1955. *Automobile Workers and the American Dream*. Garden City, N.Y.: Doubleday.

Chudacoff, Howard P. 1989. *How Old Are You? Age Consciousness in American Culture*. Princeton: Princeton University Press.

Clark, Burton R. 1987a. *The Academic Life: Small Worlds, Different Worlds.* Princeton: Carnegie Foundation for the Advancement of Teaching.

―――. 1977. *Academic Power in Italy: Bureacracy and Oligarchy in a National University System.* Chicago: University of Chicago Press.

―――(ed.). 1987b. *The Academic Profession: National, Disciplinary, and Institutional Settings.* Berkeley: University of California Press.

―――. 1985. *The School and the University: An International Perspective.* Berkeley: University of California Press.

―――. 1984. *Perspectives on Higher Education: Eight Disciplinary and Comparative Views.* Berkeley: University of California Press.

Clark, Terry N. 1968. "Institutionalization of Innovations in Higher Education: Four Models." *Administrative Science Quarterly* 13:1–25.

Clarke, Adele. 1991. "Social Worlds/Arenas as Organizational Theory." In D. Maines (ed.), *Social Organization and Social Process: Essays in Honor of Anselm Strauss,* 119–158. Hawthorne, N.Y.: Aldine de Gruyter.

―――. 1990. "A Social Worlds Adventure: The Case of Reproductive Science." In S. Cozzens and Thomas Gieryn (eds.), *Theories of Science and Society,* 23–50. Bloomington: Indiana University Press.

Clarke, Adele, and Joan H. Fujimura (eds.). 1992. *The Right Tools for the Job: At Work in Twentieth-Century Life Sciences.* Princeton: Princeton University Press.

Clausen, John A. 1993. *American Lives: Looking Back at Children of the Great Depression.* New York: Free Press.

Clifford, Geraldine Joncich, and James W. Guthrie. 1988. *Ed School.* Chicago: University of Chicago Press.

Cogan, Morris L. 1953. "Toward a Definition of Profession." *Harvard Educational Review* 23:33–50.

Cohler, Bertram J. 1982. "Personal Narrative and Life Course." In Paul B. Baltes and Orville G. Brim, Jr. (eds.), *Life-Span Development and Behavior,* vol. 4, 205–241. New York: Academic Press.

Cole, Jonathan R. 1979. *Fair Science: Women in the Scientific Community.* New York: Columbia University Press.

Cole, Jonathan R., and Stephen Cole. 1973. *Social Stratification in Science.* Chicago: University of Chicago Press.

Cole, Jonathan R., and Harriet Zuckerman. 1987. "Marriage, Motherhood and Research Performance in Science." *Scientific American* 256:119–125.

―――. 1984. "The Productivity Puzzle: Persistence and Change in Patterns of Publication of Men and Women Scientists." In *Advances in Motivation and Achievement,* vol. 2, 217–258. Greenwich, Conn.: JAI Press.

Cole, Stephen. 1992. *Making Science: Between Nature and Society.* Cambridge: Harvard University Press.

―――. 1979. "Age and Scientific Performance." *American Journal of Sociology* 84:958–977.

―――. 1970. "Professional Standing and the Reception of Scientific Discoveries." *American Journal of Sociology* 76:286–306.

Cole, Stephen, and Jonathan R. Cole. 1967. "Scientific Output and Recognition: A Study in the Operation of the Reward System in Science." *American Sociological Review* 32:377–390.

Cole, Thomas R. 1992. *The Journey of Life: A Cultural History of Aging in America.* Cambridge: Cambridge University Press.

Coleman, James S. 1969. Foreword to Bernard C. Rosen, Harry J. Crockett, Jr., and Clyde Z. Nunn (eds.), *Achievement in American Society.* Cambridge, Mass.: Schenkman.

Collin, Audrey, and Richard A. Young. 1986. "New Directions for Theories of Career." *Human Relations* 39:837–853.

Committee on Science, Engineering, and Public Policy. 1995. *Reshaping the Graduate Education of Scientists and Engineers.* Washington, D.C.: National Academy Press.

Cook, Gary A. 1993. *George Herbert Mead: The Making of a Social Pragmatist.* Urbana: University of Illinois Press.

Cookson, Peter W., and Caroline Hodges Persell. 1985. *Preparing for Power: America's Elite Boarding Schools.* New York: Basic.

Cooley, Charles Horton. 1909. *Social Organization: A Study of the Larger Mind.* New York: Charles Scribner's Sons.

———. 1902. *Human Nature and the Social Order.* New York: Charles Scribner's Sons.

Coser, Lewis A. 1971. *Masters of Sociological Thought: Ideas in Historical and Social Context.* New York: Harcourt Brace Jovanovich.

Crane, Diana. 1970. "The Academic Marketplace Revisited: A Study of Faculty Mobility Using the Cartter Ratings." *American Journal of Sociology* 75:953–964.

———. 1969. "Social Class Origin and Academic Success: The Influence of Two Stratification Systems on Academic Careers." *Sociology of Education* 42:1–17.

———. 1965. "Scientists at Major and Minor Universities: A Study of Productivity and Recognition." *American Sociological Review* 30:699–714.

Czarniawska, Barbara. 1997. *Narrating the Organization: Dramas of Institutional Identity.* Chicago: University of Chicago Press.

Dahrendorf, Ralf. 1979. *Life Chances.* Chicago: University of Chicago Press.

Dannefer, Dale. 1984a. "Adult Development and Social Theory: A Paradigmatic Reappraisal." *American Sociological Review* 49:100–116.

———. 1984b. "The Role of the Social in Life-Span Developmental Psychology, Past and Future: Rejoinder to Baltes and Nesselroade." *American Sociological Review* 49: 847–850.

Darwin, Charles. [1958]. *The Autobiography of Charles Darwin, 1809–1882.* Edited by Nora Barlow. London: Collins.

Demo, David H. 1992. "The Self-Concept over Time: Research Issues and Directions." *Annual Review of Sociology* 18:303–326.

DeNora, Tia. 1995. *Beethoven and the Construction of Genius: Musical Politics in Vienna, 1792–1803.* Berkeley: University of California Press.

Denzin, Norman K. 1989. *Interpretive Biography.* Newbury Park, Calif.: Sage.

Durkheim, Emile. [1897] 1951. *Suicide.* Translated by John A. Spaulding and George Simpson. New York: Free Press.

Easterlin, Richard A. 1980. *Birth and Fortune: The Impact of Numbers on Personal Welfare.* Chicago: University of Chicago Press.

Eiduson, Bernice T. 1962. *Scientists: Their Psychological World.* New York: Basic.

Elder, Glen H., Jr. 1998. "The Life Course and Human Development." In Richard

M. Lerner (ed.), *Handbook of Child Psychology*. Vol. 1, *Theoretical Models of Human Development*, 939–991. New York: Wiley.

———. 1995. "Life Trajectories in Changing Societies." In Albert Bandura (ed.), *Self-Efficacy in Changing Societies*, 46–57. Cambridge: Cambridge University Press.

———. 1994. "Time, Human Agency, and Social Change: Perspectives on the Life Course." *Social Psychology Quarterly* 57:4–15.

———. 1993. Foreword to John A. Clausen, *American Lives: Looking Back at Children of the Great Depression*. New York: Free Press.

———. 1992. "Life Course." In Edgar F. Borgatta and Marie L. Borgatta (eds.), *Encyclopedia of Sociology*, 1120–1130. New York: Macmillan.

———. 1991. "Lives and Social Change." In Walter R. Heinz (ed.), *Theoretical Advances in Life Course Research*, 58–86. Weinheim, Germany: Deutscher Studien Verlag.

———. 1985. "Perspectives on the Life Course." In *Life Course Dynamics: Trajectories and Transitions, 1968–1980*, 23–49. Ithaca: Cornell University Press.

———. 1981. "History and the Life Course." In Daniel Bertaux (ed.), *Biography and Society: The Life History Approach in the Social Sciences*, 77–115. Newbury Park, Calif.: Sage.

———. 1979. "Historical Change in Life Patterns and Personality." In Paul B. Baltes and Orville G. Brim (eds.), *Life-Span Development and Behavior*, vol. 2, 117–159. New York: Academic Press.

———. 1975. "Age Differentiation and the Life Course." *Annual Review of Sociology* 1:165–190.

———. 1974. *Children of the Great Depression*. Chicago: University of Chicago Press.

Elder, Glen H., Jr., and Rand D. Conger. Forthcoming. *New Worlds, New Lives: Rural Generations at Century's End*.

Elder, Glen H., Jr., Linda K. George, and Michael J. Shanahan. 1996. "Psychosocial Stress over the Life Course." In Howard B. Kaplan (ed.), *Psychosocial Stress: Perspectives on Structure, Theory, Life Course, and Methods*, 247–292. Orlando, Fla.: Academic Press.

Elder, Glen H., Jr., John Modell, and Ross D. Parke. 1993. *Children in Time and Place: Developmental and Historical Insights*. Cambridge: Cambridge University Press.

Elder, Glen H., Jr., and Angela M. O'Rand. 1995. "Adult Lives in a Changing Society." Chap. 17 in Karen S. Cook, Gary Alan Fine, and James S. House (eds.), *Sociological Perspectives on Social Psychology*, 452–475. Boston: Allyn and Bacon.

Elder, Glen H., Jr., Laura Rudkin, and Rand D. Conger. 1994. "Intergenerational Continuity and Change in Rural America." In Vern L. Bengston, K. Warner Schaie, and Linda M. Burton (eds.), *Adult Intergenerational Relations: Effects of Societal Change*, 30–78. New York: Springer.

Ellis, Susanne D. 1993. "Initial Employment of Physics Doctorate Recipients: Class of 1992." *Physics Today* 46 (December): 29–33.

Epstein, Joseph (ed.). 1995. "My Friend Edward." *American Scholar* 63 (summer): 371–394.

———. 1980. *Ambition: The Secret Passion*. New York: E. P. Dutton.

Erikson, Erik H. 1982. *The Life Cycle Completed*. New York: Norton.

————. 1959. *Identity and the Life Cycle*. New York: Norton.

————. 1950. *Childhood and Society*. New York: Norton.

Etzioni, Amitai. 1969. *The Semi-Professions and Their Organization*. Glencoe, Ill.: Free Press.

Faulkner, Robert R. 1987. *Music on Demand: Composers and Careers in the Hollywood Film Industry*. New Brunswick, N.J.: Transaction.

————. 1985. *Hollywood Studio Musicians: Their Work and Careers in the Recording Industry*. New York: University Press of America.

————. 1974. "Coming of Age in Organizations: A Comparative Study of Career Contingencies and Adult Socialization." *Sociology of Work and Occupations* 1:131–173.

Featherman, David L. 1983. "The Life-Span Perspective in Social Science Research." In Paul B. Baltes and Orville G. Brim, Jr. (eds.), *Life-Span Development and Behavior*, vol. 5, 1–57. New York: Academic Press.

Fine, Gary Alan. 1996. *Kitchens: The Culture of Restaurant Work*. Berkeley: University of California Press.

Foote, Nelson N. 1951. "Identification as the Basis for a Theory of Motivation." *American Sociological Review* 16:14–22.

Fox, Mary Frank. 1996. "Women, Academia, and Careers in Science and Engineering." In C. S. Davis, A. Ginorio, C. Hollenshead, B. Lazarus, and P. Rayman (eds.), *The Equity Equation: Women in Science, Engineering, and Mathematics*, 265–289. San Francisco: Jossey-Bass.

————. 1995. "Women and Scientific Careers." In S. Jasanoff, G. E. Markle, J. C. Petersen, and T. J. Pinch (eds.), *Handbook of Science and Technology Studies*, 205–223. Thousand Oaks, Calif.: Sage.

————. 1985. "Publication, Performance, and Reward in Science and Scholarship." In J. Smart (ed.), *Higher Education: Handbook of Theory and Research*, 255–282. New York: Agathon.

————. 1983. "Publication Productivity among Scientists: A Critical Review." *Social Studies of Science* 13:285–305.

Frank, Robert H. 1985. *Choosing the Right Pond: Human Behavior and the Quest for Status*. New York: Oxford University Press.

Freidson, Eliot. 1986. *Professional Powers: A Study of the Institutionalization of Formal Knowledge*. Chicago: University of Chicago Press.

————. 1970. *Profession of Medicine: A Study of the Sociology of Applied Knowledge*. Chicago: University of Chicago Press.

Fuchs, Victor R. 1983. *How We Live: An Economic Perspective on Americans from Birth to Death*. Cambridge: Harvard University Press.

Galison, Peter. 1997. *Image and Logic: A Material Culture of Microphysics*. Chicago: University of Chicago Press.

Galison, Peter, and Bruce Hevly (eds.). 1992. *Big Science: The Growth of Large-Scale Research*. Stanford: Stanford University Press.

Garfinkel, Harold. 1967. *Studies in Ethnomethodology*. Cambridge, England: Polity Press.

Gaston, Jerry. 1978. *The Reward System in British and American Science*. New York: Wiley.

————. 1973. *Originality and Competition in Science: A Study of the British High Energy Physics Community*. Chicago: University of Chicago Press.

George, Alexander L., and Juliette L. George. 1956. *Woodrow Wilson and Colonel House: A Personality Study.* New York: John Day Company.

George, Linda K. 1993. "Sociological Perspectives on Life Transitions." *Annual Review of Sociology* 19:353–373.

Gergen, Mary M., and Kenneth J. Gergen. 1984. "The Social Construction of Narrative Accounts." In Kenneth J. Gergen and Mary M. Gergen (eds.), *Historical Social Psychology,* 173–189. Hillsdale, N.J.: Lawrence Erlbaum.

Gerson, Kathleen. 1985. *Hard Choices: How Women Decide about Work, Career, and Motherhood.* Berkeley: University of California Press.

Gieryn, Thomas F. Forthcoming. *Cultural Cartography of Science: Episodes of Boundary Work, Sociologically Rendered.* Chicago: University of Chicago Press.

———. 1983. "Boundary-Work and the Demarcation of Science from Non-Science: Strains and Interests in Professional Ideologies of Scientists." *American Sociological Review* 48:781–795.

Glaser, Barney G. 1964a. "Comparative Failure in Science." *Science* 143:1012–1014.

———. 1964b. *Organizational Scientists: Their Professional Careers.* Indianapolis: Bobbs-Merrill.

Glaser, Barney G., and Anselm L. Strauss. 1971. *Status Passage.* Chicago: Aldine.

Goffman, Erving. 1967. *Interaction Ritual.* New York: Pantheon.

———. 1963. *Stigma: Notes on the Management of Spoiled Identity.* New York: Simon and Schuster.

———. 1961. "The Moral Career of the Mental Patient." In *Asylums: Essays on the Social Situation of Mental Patients and Other Inmates,* 125–169. Garden City, N.Y.: Doubleday.

———. 1959. *The Presentation of Self in Everyday Life.* Garden City, N.Y.: Doubleday.

———. 1952. "On Cooling the Mark Out." *Psychiatry* 15:451–463.

Goldberger, Marvin L., Brendan A. Maher, and Pamela Ebert Flattau (eds.). 1995. *Research-Doctorate Programs in the United States: Continuity and Change.* Washington, D.C.: National Academy Press.

Goode, William J. 1978. *The Celebration of Heroes: Prestige as a Control System.* Berkeley: University of California Press.

Gordon, Milton. 1975. "Toward a General Theory of Racial and Ethnic Group Relations." In Nathan Glazer and Daniel Patrick Moynihan (eds.), *Ethnicity: Theory and Experience,* 84–110. Cambridge: Harvard University Press.

Gouldner, Alvin W. 1957–58. "Cosmopolitans and Locals: Toward an Analysis of Latent Social Roles I and II." *Administrative Science Quarterly* 2:281–306, 444–480.

Graebner, William. 1980. *A History of Retirement: The Meaning and Function of an American Institution, 1885–1978.* New Haven: Yale University Press.

Greenwood, Ernest. 1957. "Attributes of a Profession." *Social Work* 2:45–55.

Gustin, Bernard H. 1973. "Charisma, Recognition, and the Motivation of Scientists." *American Journal of Sociology* 78:1119–1134.

Habenstein, R. 1954. "The Career of the Funeral Director." Ph.D. diss., University of Chicago.

Haber, Samuel. 1991. *The Quest for Authority and Honor in the American Professions, 1750–1900.* Chicago: University of Chicago Press.

Hagestad, Gunhild O. 1990. "Social Perspectives on the Life Course." In Robert

H. Binstock and Linda K. George (eds.), *Handbook of Aging and the Social Sciences*, 151–168. 3d ed. San Diego: Academic Press.

Hagstrom, Warren O. 1971. "Inputs, Outputs, and the Prestige of University Science Departments." *Sociology of Education* 44:375–397.

———. 1965. *The Scientific Community.* New York: Basic.

Halliday, Terrence. 1987. *Beyond Monopoly: Lawyers, State Crises, and Professional Empowerment.* Chicago: University of Chicago Press.

Halsey, A. H., and M. A. Trow. 1971. *The British Academics.* Cambridge: Harvard University Press.

Hankins, Thomas L. 1985. *Science and the Enlightenment.* Cambridge: Cambridge University Press.

Hareven, Tamara. 1986. "Historical Changes in the Social Construction of the Life Course." *Human Development* 29:171–177.

———(ed.). 1978. *Transitions: The Family and the Life Course in Historical Perspective.* New York: Academic Press.

Hargens, Lowell L. 1975. *Patterns of Scientific Research: A Comparative Analysis of Research in Three Scientific Fields.* Washington, D.C.: American Sociological Association.

———. 1969. "Patterns of Mobility of New Ph.D.'s among American Academic Institutions." *Sociology of Education* 42:18–37.

Hargens, Lowell L., and Warren O. Hagstrom. 1967. "Sponsored and Contest Mobility of American Academic Scientists." *Sociology of Education* 40:24–38.

Hargens, Lowell L., James C. McCann, and Barbara F. Reskin. 1978. "Productivity and Reproductivity: Fertility and Professional Achievement among Research Scientists." *Social Forces* 57:154–163.

Heinich, Nathalie. 1996. *The Glory of Van Gogh: An Anthropology of Admiration.* Princeton: Princeton University Press.

Heinz, John P., and Edward O. Laumann. 1982. *Chicago Lawyers: The Social Structure of the Bar.* New York: Russell Sage Foundation.

Held, Thomas. 1986. "Institutionalization and Deinstitutionalization of the Life Course." *Human Development* 29:157–162.

Hermanowicz, Joseph C. 1996. Review of *Gender Differences in Science Careers: The Project Access Study,* by Gerhard Sonnert and Gerald Holton. *American Journal of Sociology* 101:1133–1135.

Hirsch, Walter. 1968. *Scientists in American Society.* New York: Random House.

Hochschild, Arlie Russell. 1997. *The Time Bind: When Work Becomes Home and Home Becomes Work.* New York: Henry Holt and Company.

———. 1983. *The Managed Heart: Commercialization of Human Feeling.* Berkeley: University of California Press.

Hochschild, Arlie, and Anne Machung. 1989. *The Second-Shift: Working Parents and the Revolution at Home.* New York: Viking.

Horney, Karen. 1945. *Our Inner Conflicts.* New York: W. W. Norton.

Hughes, Everett C. 1994. *On Work, Race, and the Sociological Imagination.* Edited by Lewis A. Coser. Chicago: University of Chicago Press.

———. [1971] 1993. *The Sociological Eye.* New Brunswick, N.J.: Transaction.

———. 1958. *Men and Their Work.* Glencoe, Ill.: Free Press.

Janowitz, Morris. 1975. "Sociological Theory and Social Control." *American Journal of Sociology* 81:82–108.

————. 1960. *The Professional Soldier.* New York: Free Press.

Jasanoff, S., G. E. Markle, J. C. Petersen, and T. J. Pinch (eds.). 1995. *Handbook of Science and Technology Studies.* Thousand Oaks, Calif.: Sage.

Jencks, Christopher, and David Riesman. 1968. *The Academic Revolution.* Garden City, N.Y.: Doubleday.

Johnson, Terence. 1972. *Professions and Power.* London: Macmillan.

Jones, Lyle V., Gardner Lindzey, and Porter E. Coggeshall (eds.). 1982. *An Assessment of Research-Doctorate Programs in the United States: Mathematical and Physical Sciences.* Washington, D.C.: National Academy Press.

Joravsky, Ben. 1995. *Hoop Dreams.* Atlanta: Turner.

Kagan, Jerome. 1980. "Perspectives on Continuity." In Orville G. Brim, Jr., and Jerome Kagan (eds.), *Constancy and Change in Human Development,* 26–74. Cambridge: Harvard University Press.

Kanter, Rosabeth Moss. 1989. "Careers and the Wealth of Nations: A Macro-Perspective on the Structure and Implications of Career Forms." In Michael B. Arthur, Douglas T. Hall, and Barbara S. Lawrence (eds.), *Handbook of Career Theory,* 506–521. Cambridge: Cambridge University Press.

————. 1977. *Men and Women of the Corporation.* New York: Basic.

Kaplan, Norman. 1964. "The Sociology of Science." In Robert E. L. Faris (ed.), *Handbook of Modern Sociology,* 852–881. Chicago: Rand McNally.

Keating, Peter, Alberto Cambrosio, and Michael Mackenzie. 1992. "The Tools of the Discipline: Standards, Models, and Measures in the Affinity/Avidity Controversy in Immunology." Chap. 11 in Adele E. Clarke and Joan H. Fujimura (eds.), *The Right Tools for the Job: At Work in Twentieth-Century Life Sciences,* 312–354. Princeton: Princeton University Press.

Kevles, Daniel J. 1971. *The Physicists: The History of a Scientific Community in Modern America.* Cambridge: Harvard University Press.

Kirby, Kate, and Roman Czujko. 1993. "The Physics Job Market: Bleak for Young Scientists." *Physics Today* 46 (December): 22–27.

Knorr-Cetina, Karin. 1995. "Laboratory Studies: The Cultural Approach to the Study of Science." In S. Jasanoff, G. E. Markle, J. C. Petersen, and T. J. Pinch (eds.), *Handbook of Science and Technology Studies,* 140–166. Thousand Oaks, Calif.: Sage.

Kohli, Martin. 1986. "Social Organization and Subjective Construction of the Life Course." In Aage B. Sorensen, Franz E. Weinert, and Lourie R. Sherrod (eds.), *Human Development and the Life Course: Multidisciplinary Perspectives,* 271–292. Hillsdale, N.J.: Lawrence Erlbaum.

Kohli, Martin, and John W. Meyer. 1986. "Social Structure and Social Construction of Life Stages." *Human Development* 29:145–180.

Kohn, Melvin L., and Carmi Schooler. 1983. *Work and Personality: An Inquiry into the Impact of Social Stratification.* Norwood, N.J.: Alblex.

Kornhauser, William. 1962. *Scientists in Industry: Conflict and Accommodation.* Berkeley: University of California Press.

Krieger, Martin H. 1992. *Doing Physics: How Physicists Take Hold of the World.* Bloomington: Indiana University Press.

LaFollette, Marcel C. 1990. *Making Science Our Own: Public Images of Science, 1910–1955.* Chicago: University of Chicago Press.

Lamont, Michele. 1992. *Money, Morals, and Manners: The Culture of the French and the American Upper-Middle Classes*. Chicago: University of Chicago Press.

Lankford, John. 1997. *American Astronomy: Community, Careers, and Power, 1859–1940*. Chicago: University of Chicago Press.

Larson, Magali Sarfatti. 1978. *The Rise of Professionalism*. Berkeley: University of California Press.

Latour, Bruno, and Steve Woolgar. 1979. *Laboratory Life: The Construction of Scientific Facts*. Newbury Park, Calif.: Sage.

Lazarsfeld, Paul F., and Wagner Thielens, Jr. 1958. *The Academic Mind*. Glencoe, Ill.: Free Press.

Levine, Donald N. 1995. *Visions of the Sociological Tradition*. Chicago: University of Chicago Press.

Levinson, Daniel J. 1996. *The Seasons of a Woman's Life*. New York: Alfred A. Knopf.
———. 1978. *The Seasons of a Man's Life*. New York: Ballantine.

Linde, Charlotte. 1993. *Life Stories: The Creation of Coherence*. New York: Oxford University Press.

Lindley, David. 1993. *The End of Physics*. New York: Basic.

Lipset, Seymour Martin. 1996. *American Exceptionalism*. New York: Norton.

Long, J. Scott. 1990. "The Origins of Sex Differences in Science." *Social Forces* 68:1297–1315.
———. 1978. "Productivity and Academic Position in the Scientific Career." *American Sociological Review* 43:889–908.

Long, J. Scott, Paul D. Allison, and Robert McGinnis. 1979. "Entrance into the Academic Career." *American Sociological Review* 44:816–830.

Long, J. Scott, and Mary Frank Fox. 1995. "Scientific Careers: Universalism and Particularism." *Annual Review of Sociology* 21:45–71.

Long, J. Scott, and Robert McGinnis. 1981. "Organizational Context and Scientific Productivity." *American Sociological Review* 46:422–442.

Lortie, Dan C. 1975. *Schoolteacher: A Sociological Study*. Chicago: University of Chicago Press.
———. 1959. "From Laymen to Lawmen: Law School, Careers, and Professional Socialization." *Harvard Educational Review* 29:352–369.

Ludwig, Arnold M. 1995. *The Price of Greatness: Resolving the Creativity and Madness Controversy*. New York: Guilford.

Lundberg, George A. 1947. *Can Science Save Us?* New York: Longmans, Green, and Co.

Lynch, Michael. 1993. *Scientific Practice and Ordinary Action: Ethnomethodology and Social Studies of Science*. Cambridge: Cambridge University Press.

Maines, David R. 1992. "Life Histories." In Edgar Borgatta and Marie L. Borgatta (eds.), *Encyclopedia of Sociology*, 1134–1138. New York: Macmillan.

Mannheim, Karl. 1952. "The Problem of Generations." In *Essays on the Sociology of Knowledge*, 276–320. London: Routledge and Kegan Paul.

Marcson, Simon. 1960. *The Scientist in American Industry*. Princeton: Princeton University Press.

Marshall, T. H. 1939. "The Recent History of Professionalism in Relation to Social Structure and Social Policy." *Canadian Journal of Economics and Political Science* 5:325–340.

Marshall, Victor W. 1980. *Last Chapters: A Sociology of Aging and Dying*. Monterey, Calif.: Brooks, Cole.

Mayer, Karl Ulrich. 1986a. "The State and the Structure of the Life Course." In Aage B. Sorensen, Franz E. Weinert, and Lourie R. Sherrod (eds.), *Human Development and the Life Course: Multidisciplinary Perspectives*, 217–245. Hillsdale, N.J.: Lawrence Erlbaum.

———. 1986b. "Structural Constraints on the Life Course." *Human Development* 29:163–170.

Mayer, Karl Ulrich, and Urs Schoepflin. 1989. "The State and the Life Course." *Annual Review of Sociology* 15:187–209.

McAllister, James W. 1996. *Beauty and Revolution in Science*. Ithaca: Cornell University Press.

McClelland, David C. 1961. *The Achieving Society*. New York: Free Press.

McHugh, Peter. 1968. *Defining the Situation: The Organization of Meaning in Social Interaction*. Indianapolis: Bobbs-Merrill.

Mead, George Herbert. 1934. *Mind, Self, and Society*. Edited by Charles W. Morris. Chicago: University of Chicago Press.

———. 1932. *The Philosophy of the Present*. Edited by Arthur E. Murphy. Chicago: University of Chicago Press.

Merton, Robert K. 1973a. "The Matthew Effect in Science." In *The Sociology of Science: Theoretical and Empirical Investigations*, 439–459. Chicago: University of Chicago Press. Article first published in 1968.

———. 1973b. "The Normative Structure of Science." In *The Sociology of Science: Theoretical and Empirical Investigations*, 267–278. Chicago: University of Chicago Press. Article first published in 1942.

———. 1973c. "Priorities in Scientific Discovery." In *The Sociology of Science: Theoretical and Empirical Investigations*, 286–324. Chicago: University of Chicago Press. Article first published in 1957.

———. 1973d. " 'Recognition' and 'Excellence': Instructive Ambiguities." In *The Sociology of Science: Theoretical and Empirical Investigations*, 419–438. Chicago: University of Chicago Press. Article first published in 1960.

———. 1957. "Patterns of Influence: Local and Cosmopolitan Influentials." In *Social Theory and Social Structure*, 441–474. Glencoe, Ill.: Free Press.

Merton, Robert K., George Reader, and Patricia Kendall (eds.) 1957. *The Student-Physician*. Cambridge: Harvard University Press.

Merton, Robert K., and Harriet Zuckerman. 1973. "Age, Aging, and Age Structure in Science." In Robert K. Merton, *The Sociology of Science: Theoretical and Empirical Investigations*, 497–559. Chicago: University of Chicago Press.

Meyer, John W. 1988. "Levels of Analysis: The Life Course as a Cultural Construction." In Matilda White Riley (ed.), *Social Structures and Human Lives*, 49–62. Newbury Park, Calif.: Sage.

———. 1986a. "Myths of Socialization and Personality." In T. Heller, M. Sosna, and D. Wellbery (eds.), *Reconstructing Individualism*, 208–221. Stanford: Stanford University Press.

———. 1986b. "Self and Life Course: Institutionalizaton and Its Effects." In Aage B. Sorensen, Franz E. Weinert, and Lourie R. Sherrod (eds.), *Human Development and the Life Course: Multidisciplinary Perspectives*, 199–216. Hillsdale, N.J.: Lawrence Erlbaum.

———. 1970. "The Charter: Conditions of Diffuse Socialization." In W. Richard Scott (ed.), *Social Processes and Social Structures*, 564–578. New York: Holt, Rinehart and Winston.

Meyer, John W., John Boli, George M. Thomas, and Francisco O. Ramirez. 1997. "World Society and the Nation-State." *American Journal of Sociology* 102:144–181.

Miller, Stephen J. 1970. *Prescription for Leadership: Training for the Medical Elite*. Chicago: Aldine.

Mills, C. Wright. 1940. "Situated Actions and Vocabularies of Motive." *American Sociological Review* 5:904–913.

Mitchell, W. J. T. (ed.). 1980. *On Narrative*. Chicago: University of Chicago Press.

Mitman, Gregg, and Anne Fausto-Sterling. 1992. "Whatever Happened to Planaria? C. M. Child and the Physiology of Inheritance." Chap. 6 in Adele E. Clarke and Joan H. Fujimura (eds.), *The Right Tools for the Job: At Work in Twentieth-Century Life Sciences*, 172–197. Princeton: Princeton University Press.

Modell, John. 1989. *Into One's Own: From Youth to Adulthood in the United States, 1920–1975*. Berkeley: University of California Press.

Moen, Phyllis, Glen H. Elder, Jr., and Kurt Luscher. 1995. *Examining Lives in Context: Perspectives on the Ecology of Human Development*. Washington, D.C.: American Psychological Association.

National Research Council. 1991. *Ending Mandatory Retirement for Tenured Faculty: The Consequences for Higher Education*. Washington, D.C.: National Academy Press.

National Science Board. 1996. *Science and Engineering Indicators, 1996*. Washington, D.C.: U.S. Government Printing Office.

———. 1993. *Science and Engineering Indicators, 1993*. Washington, D.C.: U.S. Government Printing Office.

———. 1987. *Science and Engineering Indicators, 1987*. Washington, D.C.: U.S. Government Printing Office.

———. 1985. *Science Indicators*. Washington, D.C.: U.S. Government Printing Office.

———. 1983. *Science Indicators*. Washington, D.C.: U.S. Government Printing Office.

———. 1981. *Science Indicators*. Washington, D.C.: U.S. Government Printing Office.

———. 1979. *Science Indicators*. Washington, D.C.: U.S. Government Printing Office.

———. 1973. *Science Indicators*. Washington, D.C.: U.S. Government Printing Office.

National Science Foundation. 1993. *Characteristics of Doctoral Scientists and Engineers in the United States: 1993*. NSF 96-302. Washington D.C.: National Science Foundation.

———. 1990. *Women and Minorities in Science and Engineering*. NSF 90-301. Washington, D.C.: National Science Foundation.

Neugarten, Bernice L. 1996. *The Meanings of Age: Selected Papers of Bernice L. Neugarten*. Edited by Dail A. Neugarten. Chicago: University of Chicago Press.

———. 1979a. "Time, Age, and the Life Cycle." *American Journal of Psychiatry* 136:887–894.

———. 1979b. "The Young-Old and the Age-Irrelevant Society." Proceedings of Couchinching Institute, Toronto, Canada. Reprinted in Bernice L. Neugarten, *The Meanings of Age: Selected Papers of Bernice L. Neugarten*, edited by Dail A. Neugarten, 47–55 (Chicago: University of Chicago Press, 1996).

———. 1974. "Age Groups in American Society and the Rise of the Young-Old." *Annals of the American Academy of Political and Social Sciences* 415:187–198.

———(ed.). 1968. *Middle Age and Aging: A Reader in Social Psychology.* Chicago: University of Chicago Press.

Neugarten, Bernice L., and Nancy Datan. 1973. "Sociological Perspetives on the Life Cycle." In Paul B. Baltes and K. Warner Schaie (eds.), *Life-Span Developmental Psychology: Personality and Socialization*, 53–79. New York: Academic Press.

Neugarten, Bernice L., Joan W. Moore, and John C. Lowe. 1965. "Age Norms, Age Constraints, and Adult Socialization." *American Journal of Sociology* 70:710–717.

Nippert-Eng, Christena E. 1996. *Home and Work: Negotiating Boundaries through Everyday Life.* Chicago: University of Chicago Press.

O'Brien, Tim. 1994. *In the Lake of the Woods.* New York: Penguin.

O'Rand, Angela M., and Margaret L. Krecker. 1990. "Concepts of the Life Cycle: Their History, Meanings, and Uses in the Social Sciences." *Annual Review of Sociology* 16:241–262.

Orbuch, Terri L. 1997. "People's Accounts Count: The Sociology of Accounts." *Annual Review of Sociology* 23:455–478.

Osherson, Samuel. 1980. *Holding On or Letting Go: Men and Career Change at Midlife.* New York: Free Press.

Parsons, Talcott. 1951. "Deviant Behavior and the Mechanisms of Social Control." Chap. 7 in *The Social System.* Glencoe, Ill.: Free Press.

———. 1949. "The Professions and Social Structure." In *Essays in Sociological Theory*, 34–49. Glencoe, Ill.: Free Press.

Parsons, Talcott, and Gerald M. Platt. 1973. *The American University.* Cambridge: Harvard University Press.

Pelz, Donald C., and Frank M. Andrews. 1966. *Scientists in Organizations: Productive Climates for Research and Development.* New York: Wiley.

Pickering, Andrew. 1995. *The Mangle of Practice: Time, Agency, and Science.* Chicago: University of Chicago Press.

———. 1984. *Constructing Quarks: A Sociological History of Particle Physics.* Edinburgh: Edinburgh University Press.

———(ed.). 1992. *Science as Practice and Culture.* Chicago: University of Chicago Press.

Plummer, Ken. 1983. *Documents of Life.* London: Allen and Unwin.

Powell, Walter W., and Paul J. DiMaggio (eds.). 1991. *The New Institutionalism in Organizational Analysis.* Chicago: University of Chicago Press.

Reif, Fred, and Anselm Strauss. 1965. "The Impact of Rapid Discovery upon the Scientist's Career." *Social Problems* 12:297–311.

Reimer, J. 1977. "Becoming a Journeyman Electrician." *Sociology of Work and Occupations* 4:87–98.

Reskin, Barbara. 1979. "Academic Sponsorship and Scientists' Careers." *Sociology of Education* 52:129–146.

———. 1978. "Scientific Productivity, Sex, and Location in the Institution of Science." *American Journal of Sociology* 83:1235–1243.

————. 1977. "Scientific Productivity and the Reward Structure of Science." *American Sociological Review* 42:491–504.

Richardson, Laurel. 1990. "Narrative and Sociology." *Journal of Contemporary Ethnography* 19:116–135.

Ricoeur, Paul. 1983–85. *Time and Narrative*, vols. 1–3. Chicago: University of Chicago Press.

Riesman, David. 1956. *Constraint and Variety in American Education*. Garden City, N.Y.: Doubleday.

Riessman, Catherine Kohler. 1993. *Narrative Analysis*. Newbury Park Calif.: Sage.

Riley, Matilda White. 1988. "On the Significance of Age in Sociology." In *Social Structures and Human Lives*, 24–45. Newbury Park, Calif.: Sage.

————. 1986. "The Dynamisms of Life Stages: Roles, People, and Age." *Human Development* 29:150–156.

Riley, Matilda White, Anne Foner, and Joan Waring. 1988. "Sociology of Age." In Neil J. Smelser (ed.), *Handbook of Sociology*, 243–290. Newbury Park, Calif.: Sage.

Riley, Matilda White, Marilyn Johnson, and Anne Foner (eds.). 1972. *Aging and Society*. Vol. 3, *A Sociology of Age Stratification*. New York: Russell Sage Foundation.

Roe, Anne. 1952. *The Making of a Scientist*. New York: Dodd, Mead.

Rosenberg, Morris. 1979. *Conceiving the Self*. New York: Basic.

Rossiter, Margaret W. 1995. *Women Scientists in America: Before Affirmative Action, 1940–1972*. Baltimore: Johns Hopkins University Press.

————. 1982. *Women Scientists in America: Struggles and Strategies to 1940*. Baltimore: Johns Hopkins University Press.

Roth, Julius A. 1963. *Timetables: Structuring the Passage of Time in Hospital Treatment and Other Careers*. Indianapolis: Bobbs-Merrill.

Roush, Wade. 1997. "Biology Departments Restructure." *Science* 275:1156–1158.

Rubin, David C. (ed.). 1986. *Autobiographical Memory*. New York: Cambridge University Press.

Ryder, Norman B. 1965. "The Cohort as a Concept in the Study of Social Change." *American Sociological Review* 30:843–861.

Sampson, Robert J., and John H. Laub. 1993. *Crime in the Making: Pathways and Turning Points through Life*. Cambridge: Harvard University Press.

Schank, R. C., and R. P. Abelson. 1977. *Scripts, Plans, Goals, and Understanding*. Hillsdale, N.J.: Lawrence Erlbaum.

Schmitt, Raymond L., and Wilbert M. Leonard II. 1986. "Immortalizing the Self through Sport." *American Journal of Sociology* 91:1088–1111.

Schwartz, Barry. 1985. "Emerson, Cooley, and the American Heroic Vision." *Symbolic Interaction* 8:103–120.

————. 1981. *Vertical Classification: A Study in Structuralism and the Sociology of Knowledge*. Chicago: University of Chicago Press.

Scott, Marvin B., and Stanford M. Lyman. 1968. "Accounts." *American Sociological Review* 33:46–62.

Segal, Erich. 1995. *Prizes*. New York: Fawcett Columbine.

————. 1985. *The Class*. New York: Bantam.

Settersten, Richard A., Jr., and Karl Ulrich Mayer. 1997. "The Measurement of Age, Age Structuring, and Life Course." *Annual Review of Sociology* 23:233–261.

Shanahan, Michael J., Glen H. Elder, Jr., and Richard A. Miech. 1997. "History and Agency in Men's Lives: Pathways to Achievement in Cohort Perspective." *Sociology of Education* 70:54–67.

Shapin, Steven. 1996. *The Scientific Revolution.* Chicago: University of Chicago Press.

———. 1995. "Here and Everywhere: Sociology of Scientific Knowledge." *Annual Review of Sociology* 21:289–321.

Shaw, Clifford R. 1930. *The Jack-Roller: A Delinquent Boy's Own Story.* Chicago: University of Chicago Press.

Sheehy, Gail. 1976. *Passages: Predictable Crises of Adult Life.* New York: Bantam.

Shils, Edward. 1983. *The Academic Ethic.* Chicago: University of Chicago Press.

Simonton, Dean Keith. 1994. *Greatness: Who Makes History and Why.* New York: Guilford.

———. 1988. *Scientific Genius: A Psychology of Science.* Cambridge: Cambridge University Press.

Snow, David A., and Leon Anderson. 1993. *Down on Their Luck: A Study of Homeless Street People.* Berkeley: University of California Press.

Sofer, Cyril. 1970. *Men at Mid-Career: A Study of British Managers and Technical Specialists.* Cambridge: Cambridge University Press.

Sonnert, Gerhard, and Gerald Holton. 1995. *Gender Differences in Science Careers: The Project Access Study.* New Brunswick, N.J.: Rutgers University Press.

Sorensen, Aage B., Franz E. Weinert, and Lourie R. Sherrod (eds.). 1986. *Human Development and the Life Course: Multidisciplinary Perspectives.* Hillsdale, N.J.: Lawrence Erlbaum.

Stebbins, Richard A. 1970. "Career: The Subjective Approach." *Sociological Quarterly* 11:32–49.

Storer, Norman. 1966. *The Social System of Science.* New York: Holt, Rhinehart, and Winston.

Strauss, Anselm L. 1993. *Continual Permutations of Action.* New York: Aldine de Gruyter.

———. 1978. "A Social World Perspective." In Anselm L. Strauss (ed.), *Creating Sociological Awareness,* 233–244. New Brunswick, N.J.: Transaction.

———. 1959. *Mirrors and Masks: The Search for Identity.* Glencoe, Ill.: Free Press.

Strauss, Anselm L., and Lee Rainwater. 1962. *The Professional Scientist: A Study of American Chemists.* Chicago: Aldine.

Super, Donald E. 1957. *The Psychology of Careers.* New York: Harper and Row.

Suttles, Gerald D. 10 July 1995. Written personal communication.

———. 1968. *The Social Order of the Slum: Ethnicity and Territory in the Inner City.* Chicago: University of Chicago Press.

Taubes, Gary. 1994. "Young Physicists Hear Wall Street Calling." *Science* 264:22.

———. 1986. *Nobel Dreams: Power, Deceit, and the Ultimate Experiment.* New York: Random House.

Thomas, W. I. 1923. *The Unadjusted Girl: With Cases and Standpoint for Behavioral Analysis.* Boston: Little, Brown.

Thomas, W. I., and Florian Znaniecki. [1918–20] 1958. *The Polish Peasant in Europe and America,* vols. 1–5. New York: Dover.

Traweek, Sharon. 1988. *Beamtimes and Lifetimes: The World of High Energy Physicists.* Cambridge: Harvard University Press.

Turner, Ralph. 1976. "The Real Self: From Institution to Impulse." *American Journal of Sociology* 81:989–1016.

Unruh, David. 1989. "Toward a Social Psychology of Reminiscence." In David Unruh and Gail E. Livings (eds.), *Current Perspectives on Aging and the Life Cycle*, vol. 3, 25–46. Greenwich, Conn.: JAI Press.

U.S. Department of Education. 1995. *Digest of Education Statistics.* Washington, D.C.: U.S. Government Printing Office.

———. 1970. *Digest of Education Statistics.* Washington, D.C.: U.S. Government Printing Office.

U.S. News & World Report. 1993. "Leaders of the Academic Pack." 118 (March): 76.

Vetter, B. 1987. "Women's Progress." *Mosaic* 18:2–9.

Walli, Kameshwar C. 1991. *Chandra: A Biography of S. Chandrasekhar.* Chicago: University of Chicago Press.

Weber, Max. 1946. "Science as a Vocation." In H. H. Gerth and C. Wright Mills (trans. and eds.), *From Max Weber: Essays in Sociology,* 129–156. New York: Oxford University Press. Article first published in 1919.

Weiss, Robert S. 1990. *Staying the Course: The Emotional and Social Lives of Men Who Do Well at Work.* New York: Free Press.

Wells, L. Edward, and Sheldon Stryker. 1988. "Stability and Change in Self over the Life Course." In Paul B. Baltes, David L. Featherman, and Richard M. Lerner (eds.), *Life-Span Development and Behavior,* vol. 8, 191–229. Hillsdale, N.J.: Lawrence Erlbaum.

WGBH. 1995. *The Making of a Doctor.* Boston: WGBH Educational Foundation.

White, Harrison C. 1993. *Careers and Creativity: Social Forces in the Arts.* Boulder, Colo.: Westview.

White, Hayden. 1987. *The Content of the Form: Narrative Discourse and Historical Representation.* Baltimore: Johns Hopkins University Press.

Wilensky, Harold. 1964. "The Professionalization of Everyone?" *American Journal of Sociology* 70:137–158.

Wiley, Norbert. 1995. *The Semiotic Self.* Chicago: University of Chicago Press.

Willis, Paul. 1977. *Learning to Labor: How Working Class Kids Get Working Class Jobs.* New York: Columbia University Press.

Wilson, Logan. 1942. *The Academic Man: A Study in the Sociology of a Profession.* London: Oxford University Press.

Yeats, William Butler. [1983]. "To a Friend Whose Work Has Come to Nothing." In Richard J. Finneran (ed.), *The Collected Poems of W. B. Yeats,* 109. New York: Macmillan.

Young, Richard A., and Audrey Collin (eds.). 1992. *Interpreting Career: Hermeneutical Studies of Lives in Context.* Westport, Conn.: Praeger.

Ziman, John. 1987. *Knowing Everything about Nothing: Specialization and Change in Scientific Careers.* Cambridge: Cambridge University Press.

Zuckerman, Harriet. 1989. "Accumulation of Advantage and Disadvantage: The Theory and Its Intellectual Biography." In C. Mongardini and S. Tabboni (eds.), *L'opera di R. K. Merton e la sociologia contemporeana,* 153–176. Genoa: Edizioni Culturali Internationali Genova.

———. 1988. "The Sociology of Science." In Neil J. Smelser (ed.), *Handbook of Sociology,* 511–574. Newbury Park, Calif.: Sage.

———. 1977. *Scientific Elite: Nobel Laureates in the United States.* New York: Free Press.

————. 1970. "Stratification in American Science." *Sociological Inquiry* 40:235–257.

Zuckerman, Harriet, and Jonathan R. Cole. 1975. "Women in American Science." *Minerva* 13:82–102.

Zuckerman, Harriet, Jonathan R. Cole, and John T. Bruer (eds.). 1991. *The Outer Circle: Women in the Scientific Community.* New York: Norton.

Zussman, Robert. 1985. *Mechanics of the Middle Class: Work and Politics among American Engineers.* Berkeley: University of California Press.

INDEX